MATH WORKBOOK

EXERCISES ---------- PAGES

ADDITION EXERCISES

1) $19 + 10 =$ ....

2) $10 + 10 =$ ....

3) $14 + 13 =$ ....

4) $12 + 6 =$ ....

5) $11 + 16 =$ ....

6) $12 + 1 =$ ....

7) $14 + 15 =$ ....

8) $1 + 4 =$ ....

9) $2 + 2 =$ ....

10) $14 + 11 =$ ....

11) $6 + 11 =$ ....

12) $11 + 5 =$ ....

13) $10 + 1 =$ ....

14) $16 + 12 =$ ....

15) $4 + 5 =$ ....

16) $4 + 2 =$ ....

17) $3 + 15 =$ ....

18) $15 + 13 =$ ....

19) $10 + 18 =$ ....

20) $6 + 10 =$ ....

Page 2

1) 2 + 14 = ...	2) 5 + 1 = ...	3) 6 + 11 = ...	4) 7 + 10 = ...
5) 13 + 14 = ...	6) 2 + 17 = ...	7) 4 + 4 = ...	8) 3 + 4 = ...
9) 11 + 5 = ...	10) 11 + 7 = ...	11) 4 + 10 = ...	12) 14 + 14 = ...
13) 2 + 10 = ...	14) 10 + 19 = ...	15) 17 + 10 = ...	16) 7 + 12 = ...
17) 3 + 10 = ...	18) 10 + 6 = ...	19) 13 + 4 = ...	20) 4 + 20 = ...

Page 3

1) $11 + 13$ ....

2) $4 + 12$ ....

3) $5 + 2$ ....

4) $17 + 10$ ....

5) $10 + 15$ ....

6) $13 + 12$ ....

7) $10 + 18$ ....

8) $7 + 20$ ....

9) $4 + 3$ ....

10) $11 + 12$ ....

11) $17 + 12$ ....

12) $13 + 5$ ....

13) $5 + 4$ ....

14) $7 + 10$ ....

15) $11 + 2$ ....

16) $13 + 11$ ....

17) $16 + 1$ ....

18) $15 + 3$ ....

19) $4 + 14$ ....

20) $15 + 13$ ....

1) $1 + 3 =$ ....

2) $14 + 12 =$ ....

3) $1 + 10 =$ ....

4) $16 + 11 =$ ....

5) $7 + 2 =$ ....

6) $15 + 12 =$ ....

7) $11 + 10 =$ ....

8) $2 + 6 =$ ....

9) $1 + 2 =$ ....

10) $19 + 10 =$ ....

11) $8 + 10 =$ ....

12) $3 + 3 =$ ....

13) $12 + 13 =$ ....

14) $6 + 10 =$ ....

15) $20 + 5 =$ ....

16) $13 + 10 =$ ....

17) $7 + 11 =$ ....

18) $13 + 11 =$ ....

19) $3 + 5 =$ ....

20) $3 + 6 =$ ....

1) $7 + 1 =$ ....

2) $2 + 14 =$ ....

3) $11 + 10 =$ ....

4) $6 + 12 =$ ....

5) $15 + 11 =$ ....

6) $2 + 15 =$ ....

7) $10 + 5 =$ ....

8) $10 + 3 =$ ....

9) $11 + 1 =$ ....

10) $2 + 13 =$ ....

11) $16 + 2 =$ ....

12) $13 + 16 =$ ....

13) $1 + 4 =$ ....

14) $3 + 16 =$ ....

15) $2 + 12 =$ ....

16) $7 + 11 =$ ....

17) $10 + 7 =$ ....

18) $11 + 5 =$ ....

19) $10 + 12 =$ ....

20) $19 + 20 =$ ....

1) $3 + 3 =$ ....

2) $4 + 1 =$ ....

3) $20 + 9 =$ ....

4) $17 + 2 =$ ....

5) $2 + 3 =$ ....

6) $15 + 11 =$ ....

7) $2 + 15 =$ ....

8) $10 + 5 =$ ....

9) $13 + 13 =$ ....

10) $1 + 1 =$ ....

11) $7 + 12 =$ ....

12) $15 + 4 =$ ....

13) $11 + 16 =$ ....

14) $19 + 10 =$ ....

15) $10 + 17 =$ ....

16) $12 + 12 =$ ....

17) $13 + 20 =$ ....

18) $4 + 15 =$ ....

19) $16 + 2 =$ ....

20) $15 + 13 =$ ....

Page 7

1) $3 + 15 = \square$

2) $3 + 6 = \square$

3) $14 + 1 = \square$

4) $13 + \square = 28$

5) $\square + 12 = 28$

6) $3 + 14 = \square$

7) $5 + 12 = \square$

8) $11 + 2 = \square$

9) $\square + 1 = 2$

10) $\square + 14 = 18$

11) $\square + 11 = 12$

12) $\square + 10 = 13$

13) $2 + \square = 19$

14) $\square + 7 = 9$

15) $4 + \square = 14$

16) $\square + 6 = 17$

17) $\square + 13 = 27$

18) $\square + 5 = 9$

19) $14 + \square = 16$

20) $15 + \square = 16$

1) $11 + 7 = \square$

2) $11 + 2 = \square$

3) $\square + 12 = 17$

4) $\square + 10 = 29$

5) $17 + 11 = \square$

6) $13 + 13 = \square$

7) $\square + 13 = 17$

8) $10 + 18 = \square$

9) $16 + 10 = \square$

10) $\square + 11 = 18$

11) $8 + 11 = \square$

12) $3 + 15 = \square$

13) $10 + 13 = \square$

14) $\square + 2 = 19$

15) $11 + \square = 25$

16) $8 + 10 = \square$

17) $15 + 3 = \square$

18) $15 + \square = 19$

19) $\square + 2 = 6$

20) $\square + 10 = 28$

Page 9

1) $1 + 13 = \square$	2) $1 + 16 = \square$	3) $14 + \square = 28$	4) $\square + 5 = 15$
5) $2 + 4 = \square$	6) $15 + \square = 28$	7) $\square + 10 = 28$	8) $\square + 5 = 16$
9) $\square + 6 = 17$	10) $4 + 14 = \square$	11) $10 + \square = 22$	12) $12 + \square = 25$
13) $11 + 7 = \square$	14) $\square + 11 = 17$	15) $8 + 10 = \square$	16) $2 + \square = 3$
17) $\square + 11 = 25$	18) $\square + 17 = 28$	19) $11 + \square = 13$	20) $13 + 5 = \square$

Page 10

1) $14 + 4 = \square$

2) $18 + \square = 28$

3) $6 + 10 = \square$

4) $\square + 10 = 26$

5) $3 + \square = 8$

6) $\square + 11 = 25$

7) $1 + \square = 9$

8) $\square + 1 = 8$

9) $11 + 15 = \square$

10) $4 + 4 = \square$

11) $3 + \square = 18$

12) $16 + \square = 36$

13) $10 + \square = 22$

14) $15 + \square = 18$

15) $2 + \square = 8$

16) $\square + 2 = 17$

17) $\square + 20 = 34$

18) $17 + \square = 28$

19) $12 + \square = 25$

20) $5 + 3 = \square$

Page 11

1) $10 + 6 = \square$

2) $15 + \square = 18$

3) $\square + 4 = 18$

4) $16 + \square = 27$

5) $1 + \square = 13$

6) $3 + 6 = \square$

7) $\square + 4 = 19$

8) $\square + 20 = 23$

9) $\square + 15 = 16$

10) $14 + 13 = \square$

11) $11 + \square = 15$

12) $12 + 12 = \square$

13) $\square + 7 = 18$

14) $\square + 13 = 28$

15) $11 + 11 = \square$

16) $3 + \square = 6$

17) $17 + \square = 29$

18) $\square + 3 = 14$

19) $\square + 20 = 25$

20) $4 + 20 = \square$

1) 9 marbles were in the basket. 7 are red and the rest are green. How many marbles are green?

2) Ellen has 8 more plums than Amy. Amy has 6 plums. How many plums does Ellen have?

3) Jake has 1 ball and Adam has 7 balls. How many balls do Jake and Adam have together?

4) Some pears were in the basket. 3 more pears were added to the basket. Now there are 5 pears. How many pears were in the basket before more pears were added?

5) 6 apples were in the basket. More apples were added to the basket. Now there are 11 apples. How many apples were added to the basket?

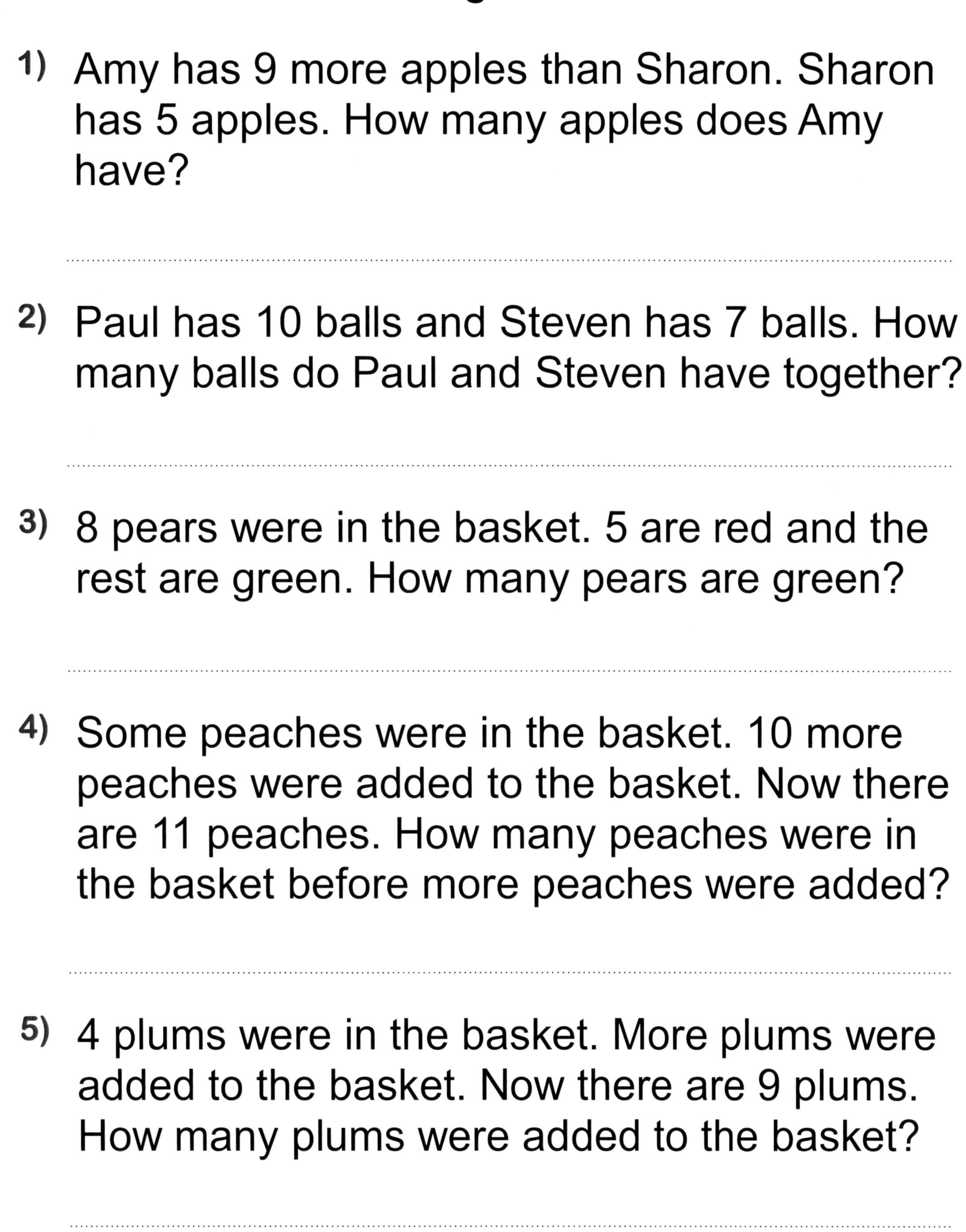

1) Amy has 9 more apples than Sharon. Sharon has 5 apples. How many apples does Amy have?

2) Paul has 10 balls and Steven has 7 balls. How many balls do Paul and Steven have together?

3) 8 pears were in the basket. 5 are red and the rest are green. How many pears are green?

4) Some peaches were in the basket. 10 more peaches were added to the basket. Now there are 11 peaches. How many peaches were in the basket before more peaches were added?

5) 4 plums were in the basket. More plums were added to the basket. Now there are 9 plums. How many plums were added to the basket?

1) 10 balls were in the basket. 9 are red and the rest are green. How many balls are green?

2) 4 red apples and 6 green apples are in the basket. How many apples are in the basket?

3) Some marbles were in the basket. 5 more marbles were added to the basket. Now there are 14 marbles. How many marbles were in the basket before more marbles were added?

4) 2 plums were in the basket. More plums were added to the basket. Now there are 4 plums. How many plums were added to the basket?

5) 5 oranges are in the basket. 9 more oranges are put in the basket. How many oranges are in the basket now?

1) Allan has 3 peaches and Brian has 8 peaches. How many peaches do Allan and Brian have together?

2) Some balls were in the basket. 7 more balls were added to the basket. Now there are 12 balls. How many balls were in the basket before more balls were added?

3) 8 marbles are in the basket. 5 more marbles are put in the basket. How many marbles are in the basket now?

4) 3 apples were in the basket. More apples were added to the basket. Now there are 9 apples. How many apples were added to the basket?

5) 12 plums were in the basket. 10 are red and the rest are green. How many plums are green?

SUBTRACTION EXERCISES

Page 16

1) $9 - 3$ ....

2) $23 - 11$ ....

3) $16 - 5$ ....

4) $9 - 6$ ....

5) $12 - 4$ ....

6) $12 - 9$ ....

7) $7 - 4$ ....

8) $27 - 12$ ....

9) $6 - 4$ ....

10) $27 - 4$ ....

11) $2 - 2$ ....

12) $12 - 3$ ....

13) $20 - 5$ ....

14) $1 - 1$ ....

15) $8 - 2$ ....

16) $11 - 7$ ....

17) $3 - 1$ ....

18) $13 - 4$ ....

19) $10 - 9$ ....

20) $21 - 17$ ....

Page 17

1) $3 - 2$ ....

2) $19 - 4$ ....

3) $11 - 2$ ....

4) $26 - 12$ ....

5) $1 - 1$ ....

6) $25 - 12$ ....

7) $15 - 11$ ....

8) $20 - 2$ ....

9) $20 - 7$ ....

10) $12 - 8$ ....

11) $11 - 4$ ....

12) $16 - 3$ ....

13) $6 - 4$ ....

14) $25 - 4$ ....

15) $4 - 3$ ....

16) $6 - 3$ ....

17) $27 - 19$ ....

18) $7 - 5$ ....

19) $4 - 2$ ....

20) $23 - 13$ ....

1) $22 - 10 =$ ....

2) $17 - 12 =$ ....

3) $19 - 6 =$ ....

4) $7 - 3 =$ ....

5) $24 - 16 =$ ....

6) $6 - 6 =$ ....

7) $25 - 20 =$ ....

8) $13 - 5 =$ ....

9) $8 - 6 =$ ....

10) $18 - 6 =$ ....

11) $15 - 15 =$ ....

12) $17 - 4 =$ ....

13) $6 - 5 =$ ....

14) $3 - 2 =$ ....

15) $6 - 2 =$ ....

16) $26 - 15 =$ ....

17) $19 - 1 =$ ....

18) $18 - 10 =$ ....

19) $1 - 1 =$ ....

20) $28 - 12 =$ ....

1) $23 - 7$ = ...

2) $11 - 3$ = ...

3) $20 - 18$ = ...

4) $12 - 7$ = ...

5) $8 - 6$ = ...

6) $2 - 1$ = ...

7) $2 - 2$ = ...

8) $4 - 2$ = ...

9) $3 - 2$ = ...

10) $11 - 4$ = ...

11) $19 - 2$ = ...

12) $5 - 1$ = ...

13) $4 - 4$ = ...

14) $11 - 8$ = ...

15) $23 - 19$ = ...

16) $17 - 3$ = ...

17) $19 - 8$ = ...

18) $16 - 5$ = ...

19) $11 - 7$ = ...

20) $7 - 4$ = ...

1) $3 - 2$ = ...

2) $21 - 4$ = ...

3) $30 - 1$ = ...

4) $13 - 8$ = ...

5) $16 - 7$ = ...

6) $9 - 4$ = ...

7) $22 - 7$ = ...

8) $2 - 2$ = ...

9) $16 - 13$ = ...

10) $12 - 5$ = ...

11) $11 - 2$ = ...

12) $26 - 18$ = ...

13) $28 - 14$ = ...

14) $29 - 16$ = ...

15) $5 - 4$ = ...

16) $6 - 2$ = ...

17) $29 - 9$ = ...

18) $13 - 3$ = ...

19) $17 - 2$ = ...

20) $15 - 12$ = ...

1) $6 - 3 =$ ....

2) $21 - 4 =$ ....

3) $13 - 5 =$ ....

4) $12 - 11 =$ ....

5) $23 - 11 =$ ....

6) $6 - 5 =$ ....

7) $2 - 2 =$ ....

8) $12 - 8 =$ ....

9) $10 - 2 =$ ....

10) $3 - 2 =$ ....

11) $19 - 14 =$ ....

12) $10 - 6 =$ ....

13) $8 - 5 =$ ....

14) $11 - 7 =$ ....

15) $5 - 2 =$ ....

16) $9 - 3 =$ ....

17) $26 - 12 =$ ....

18) $7 - 1 =$ ....

19) $12 - 9 =$ ....

20) $15 - 10 =$ ....

1) $\begin{array}{r} 11 \\ - \square \\ \hline 10 \end{array}$

2) $\begin{array}{r} 1 \\ - \square \\ \hline 0 \end{array}$

3) $\begin{array}{r} 18 \\ - \square \\ \hline 12 \end{array}$

4) $\begin{array}{r} 5 \\ - \square \\ \hline 3 \end{array}$

5) $\begin{array}{r} \square \\ - 10 \\ \hline 17 \end{array}$

6) $\begin{array}{r} \square \\ - 4 \\ \hline 3 \end{array}$

7) $\begin{array}{r} \square \\ - 1 \\ \hline 1 \end{array}$

8) $\begin{array}{r} 27 \\ - \square \\ \hline 9 \end{array}$

9) $\begin{array}{r} 30 \\ - \square \\ \hline 15 \end{array}$

10) $\begin{array}{r} \square \\ - 10 \\ \hline 14 \end{array}$

11) $\begin{array}{r} 8 \\ - \square \\ \hline 1 \end{array}$

12) $\begin{array}{r} \square \\ - 10 \\ \hline 19 \end{array}$

13) $\begin{array}{r} 16 \\ - 6 \\ \hline \square \end{array}$

14) $\begin{array}{r} 8 \\ - 5 \\ \hline \square \end{array}$

15) $\begin{array}{r} \square \\ - 9 \\ \hline 10 \end{array}$

16) $\begin{array}{r} 12 \\ - \square \\ \hline 4 \end{array}$

17) $\begin{array}{r} 21 \\ - 14 \\ \hline \square \end{array}$

18) $\begin{array}{r} 28 \\ - 19 \\ \hline \square \end{array}$

19) $\begin{array}{r} 13 \\ - \square \\ \hline 11 \end{array}$

20) $\begin{array}{r} 19 \\ - 7 \\ \hline \square \end{array}$

1) ☐ - 11 2	2) 9 - ☐ 7	3) 13 - 10 ☐	4) ☐ - 1 1
5) 24 - ☐ 12	6) 19 - ☐ 8	7) 7 - 4 ☐	8) 11 - 8 ☐
9) 22 - 13 ☐	10) 18 - 6 ☐	11) ☐ - 11 14	12) ☐ - 8 4
13) 23 - ☐ 16	14) ☐ - 9 0	15) ☐ - 14 1	16) ☐ - 3 2
17) 18 - 15 ☐	18) 3 - ☐ 0	19) 11 - ☐ 9	20) 24 - 2 ☐

Page 24

1) ☐ − 16 = 10	2) ☐ − 3 = 9	3) ☐ − 3 = 8	4) 18 − 13 = ☐
5) 12 − ☐ = 0	6) 23 − 3 = ☐	7) 1 − ☐ = 0	8) 22 − ☐ = 6
9) 30 − 14 = ☐	10) 27 − ☐ = 10	11) 4 − ☐ = 1	12) 14 − 10 = ☐
13) 16 − ☐ = 1	14) 4 − ☐ = 2	15) 19 − ☐ = 1	16) 10 − ☐ = 4
17) 16 − ☐ = 3	18) 21 − ☐ = 9	19) ☐ − 5 = 2	20) 29 − 2 = ☐

Page 25

1) $12 - 9 = \square$

2) $17 - \square = 13$

3) $22 - \square = 4$

4) $\square - 11 = 2$

5) $\square - 2 = 2$

6) $\square - 13 = 11$

7) $6 - 6 = \square$

8) $28 - \square = 24$

9) $12 - 8 = \square$

10) $\square - 9 = 10$

11) $\square - 1 = 3$

12) $20 - 5 = \square$

13) $\square - 5 = 14$

14) $\square - 1 = 2$

15) $18 - 8 = \square$

16) $22 - \square = 5$

17) $13 - 5 = \square$

18) $\square - 2 = 4$

19) $30 - \square = 17$

20) $\square - 5 = 11$

Page 26

1)	2)	3)	4)
12		11	26
- 5	- 4	-	-
	1	4	15

5)	6)	7)	8)
7	29		20
- 4	-	- 7	- 3
	18	21	

9)	10)	11)	12)
16	10		
- 8	- 3	- 19	- 8
		5	12

13)	14)	15)	16)
27	16		21
-	-	- 3	- 1
20	7	26	

17)	18)	19)	20)
28		10	27
- 17	- 15	- 10	- 2
	7		

1) 6 oranges are in the basket. 3 are red and the rest are green. How many oranges are green?

....

2) 2 plums are in the basket. 2 plums are taken out of the basket. How many plums are in the basket now?

....

3) Some balls were in the basket. 2 balls were taken from the basket. Now there is 1 ball. How many balls were in the basket before some of the balls were taken?

....

4) 4 marbles were in the basket. Some of the marbles were removed from the basket. Now there is 1 marble. How many marbles were removed from the basket?

....

5) Janet has 5 fewer pears than Michele. Michele has 10 pears. How many pears does Janet have?

....

1) Janet has 0 fewer peaches than Audrey. Audrey has 3 peaches. How many peaches does Janet have?

2) Some balls were in the basket. 3 balls were taken from the basket. Now there is 1 ball. How many balls were in the basket before some of the balls were taken?

3) 7 apples were in the basket. Some of the apples were removed from the basket. Now there are 4 apples. How many apples were removed from the basket?

4) 4 pears are in the basket. 4 are red and the rest are green. How many pears are green?

5) 4 marbles are in the basket. 4 marbles are taken out of the basket. How many marbles are in the basket now?

1) Brian has 6 balls. Allan has 9 balls. How many more balls does Allan have than Brian?

2) Michele has 0 fewer peaches than Marin. Marin has 6 peaches. How many peaches does Michele have?

3) 7 apples are in the basket. 3 are red and the rest are green. How many apples are green?

4) Some marbles were in the basket. 6 marbles were taken from the basket. Now there are 0 marbles. How many marbles were in the basket before some of the marbles were taken?

5) 5 pears were in the basket. Some of the pears were removed from the basket. Now there are 2 pears. How many pears were removed from the basket?

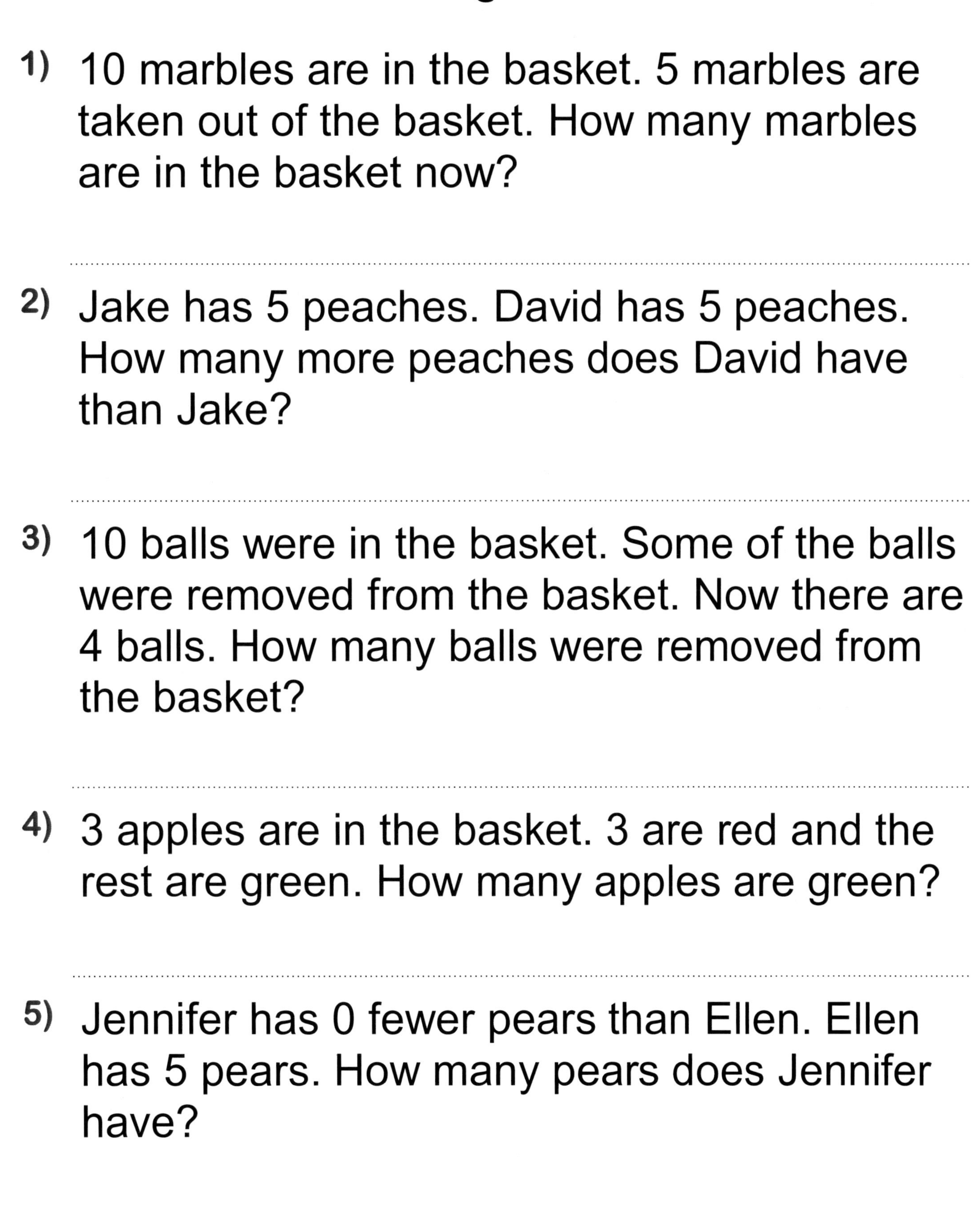

1) 10 marbles are in the basket. 5 marbles are taken out of the basket. How many marbles are in the basket now?

2) Jake has 5 peaches. David has 5 peaches. How many more peaches does David have than Jake?

3) 10 balls were in the basket. Some of the balls were removed from the basket. Now there are 4 balls. How many balls were removed from the basket?

4) 3 apples are in the basket. 3 are red and the rest are green. How many apples are green?

5) Jennifer has 0 fewer pears than Ellen. Ellen has 5 pears. How many pears does Jennifer have?

TELLING TIME EXERCISES

Page 31

1)

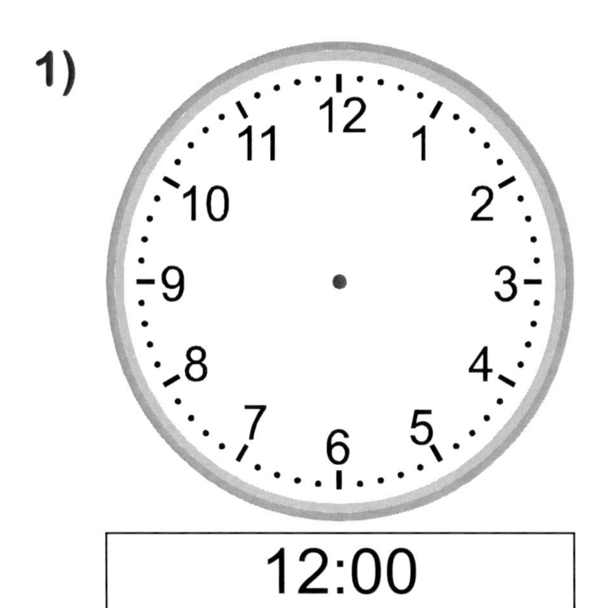

2)

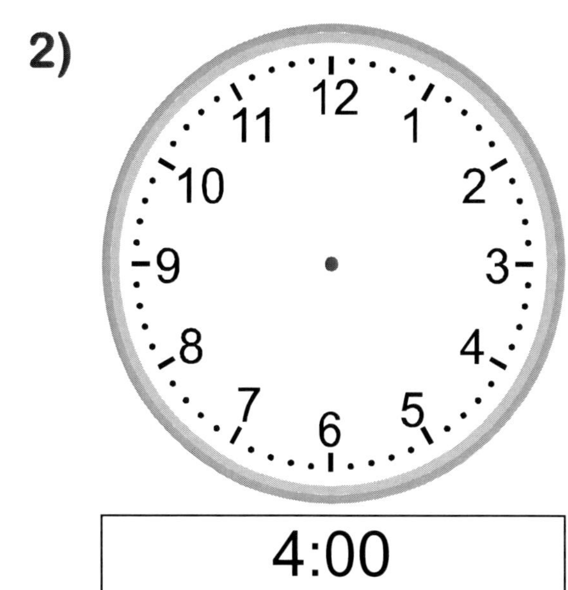

3)

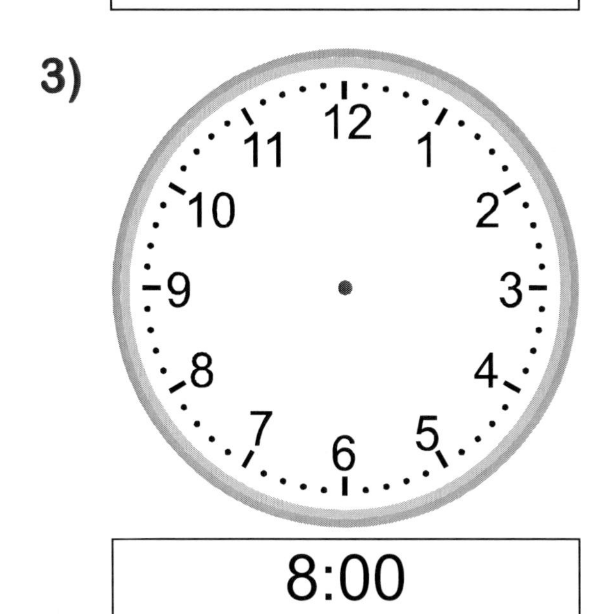

4)

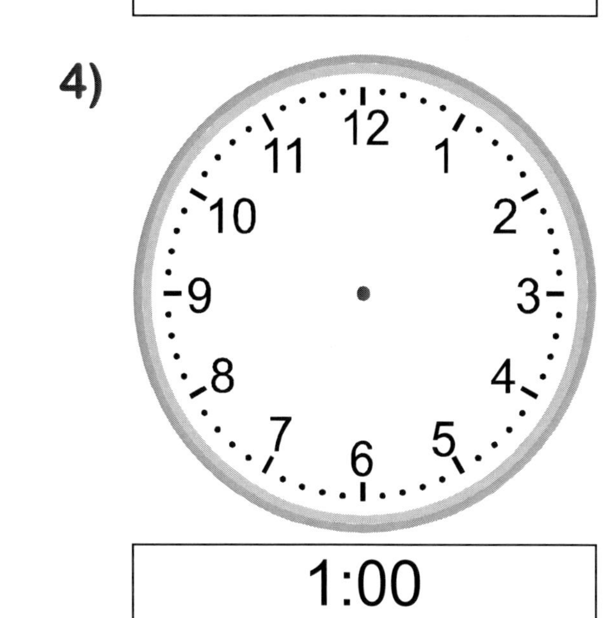

5)

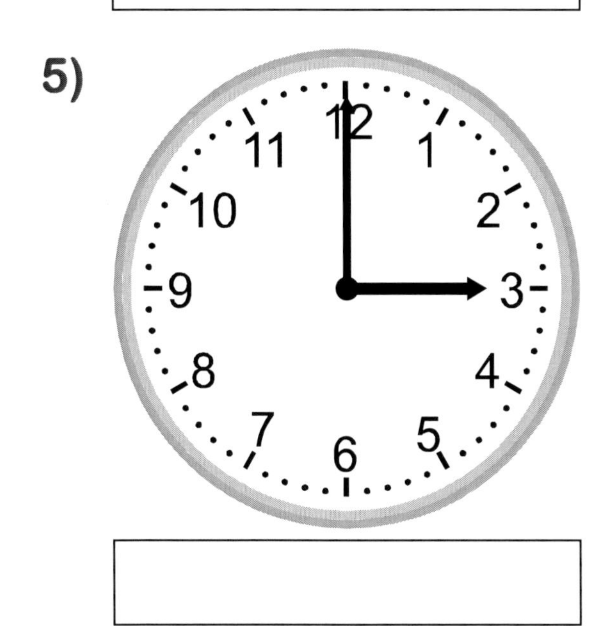

6)

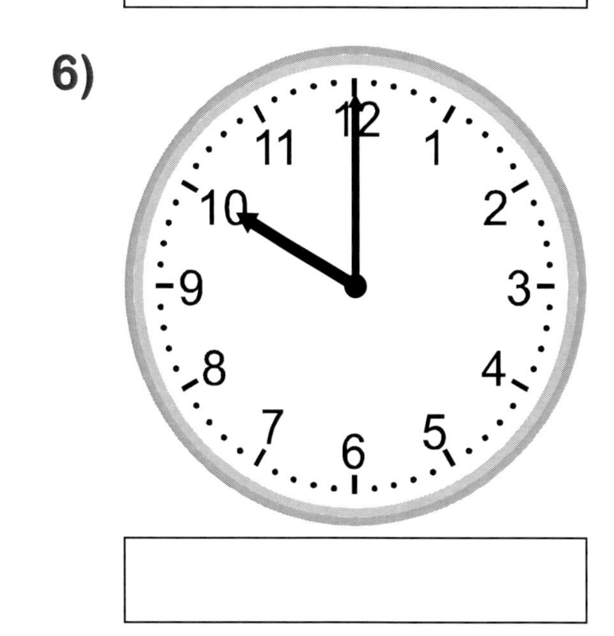

Page 32

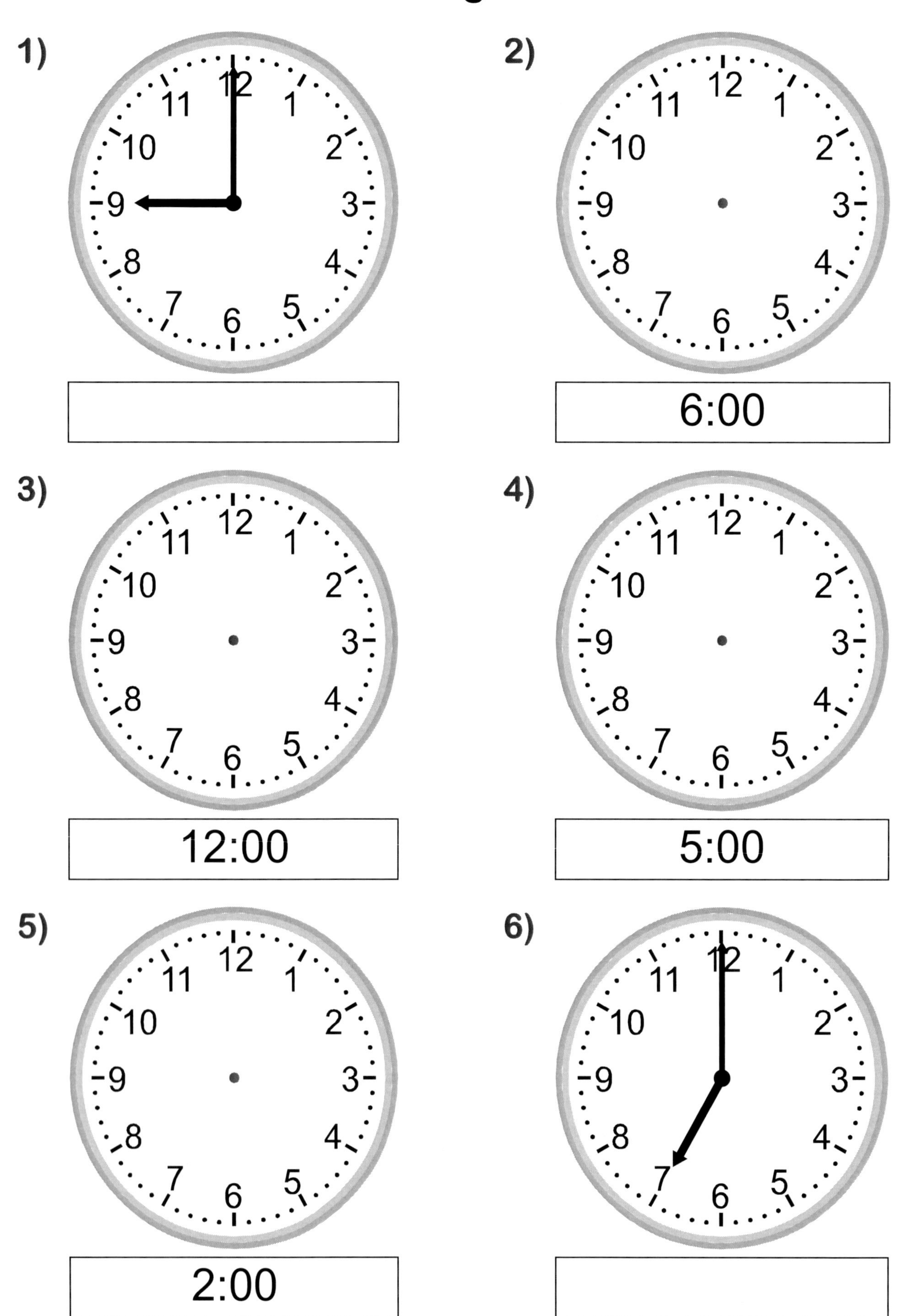

Page 33

1)

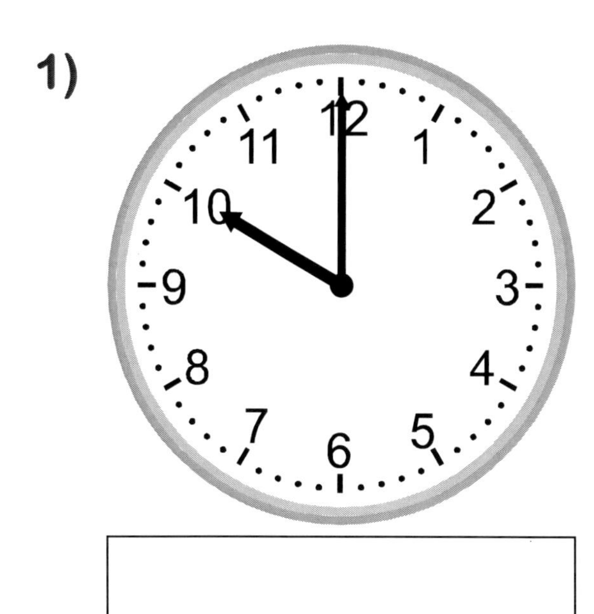

2)

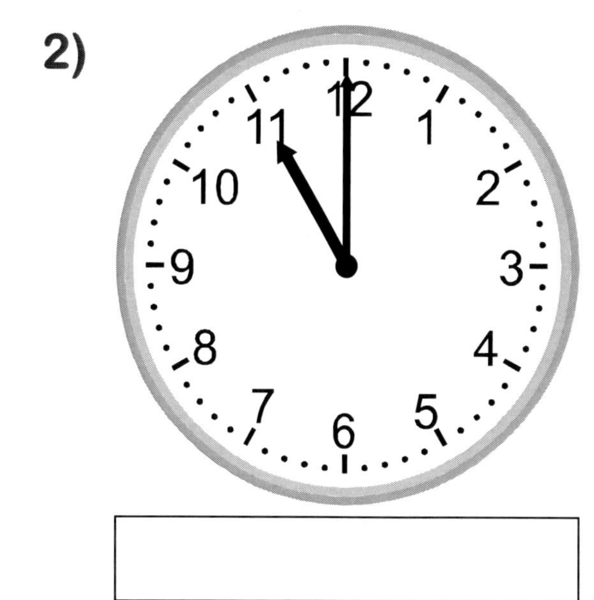

3)

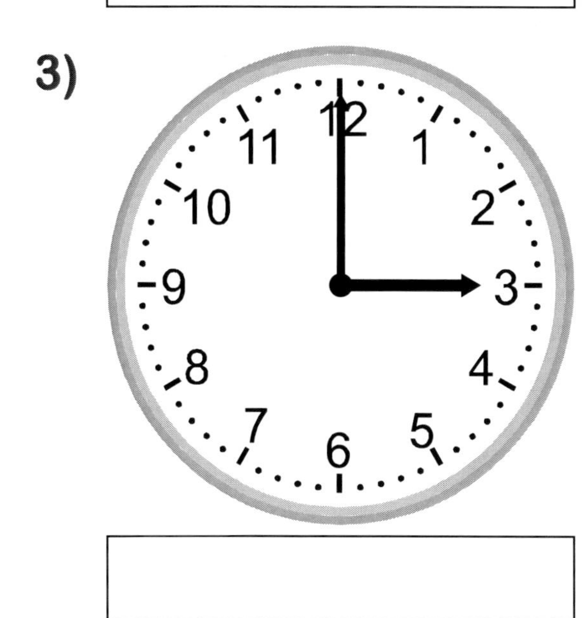

4)

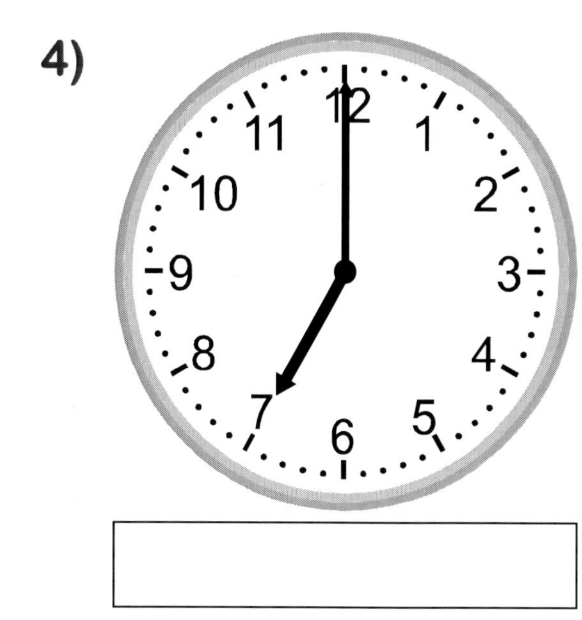

5)

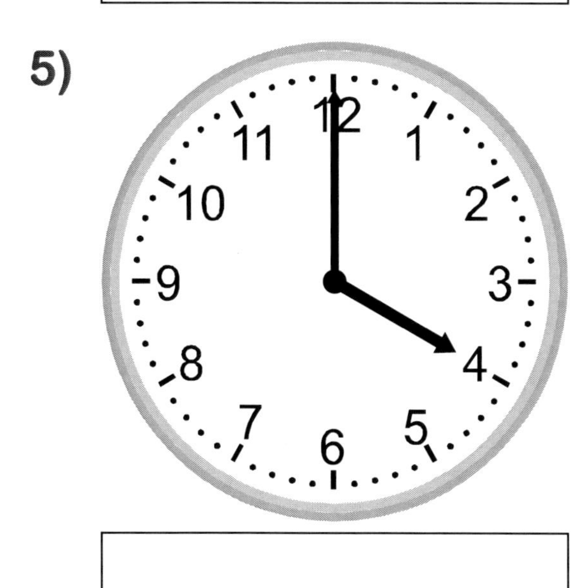

6)

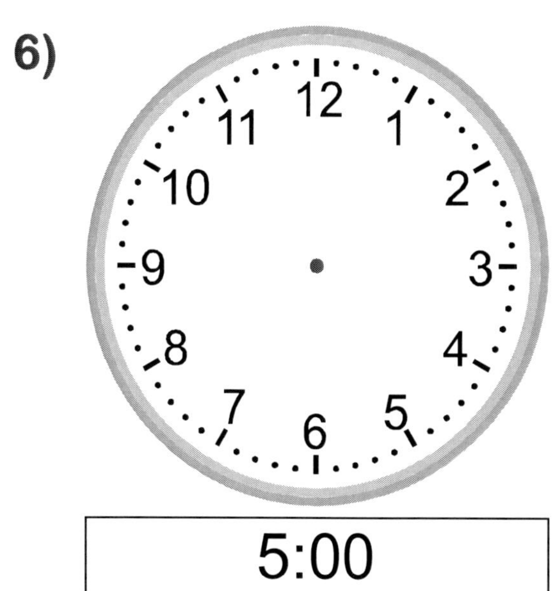

Page 34

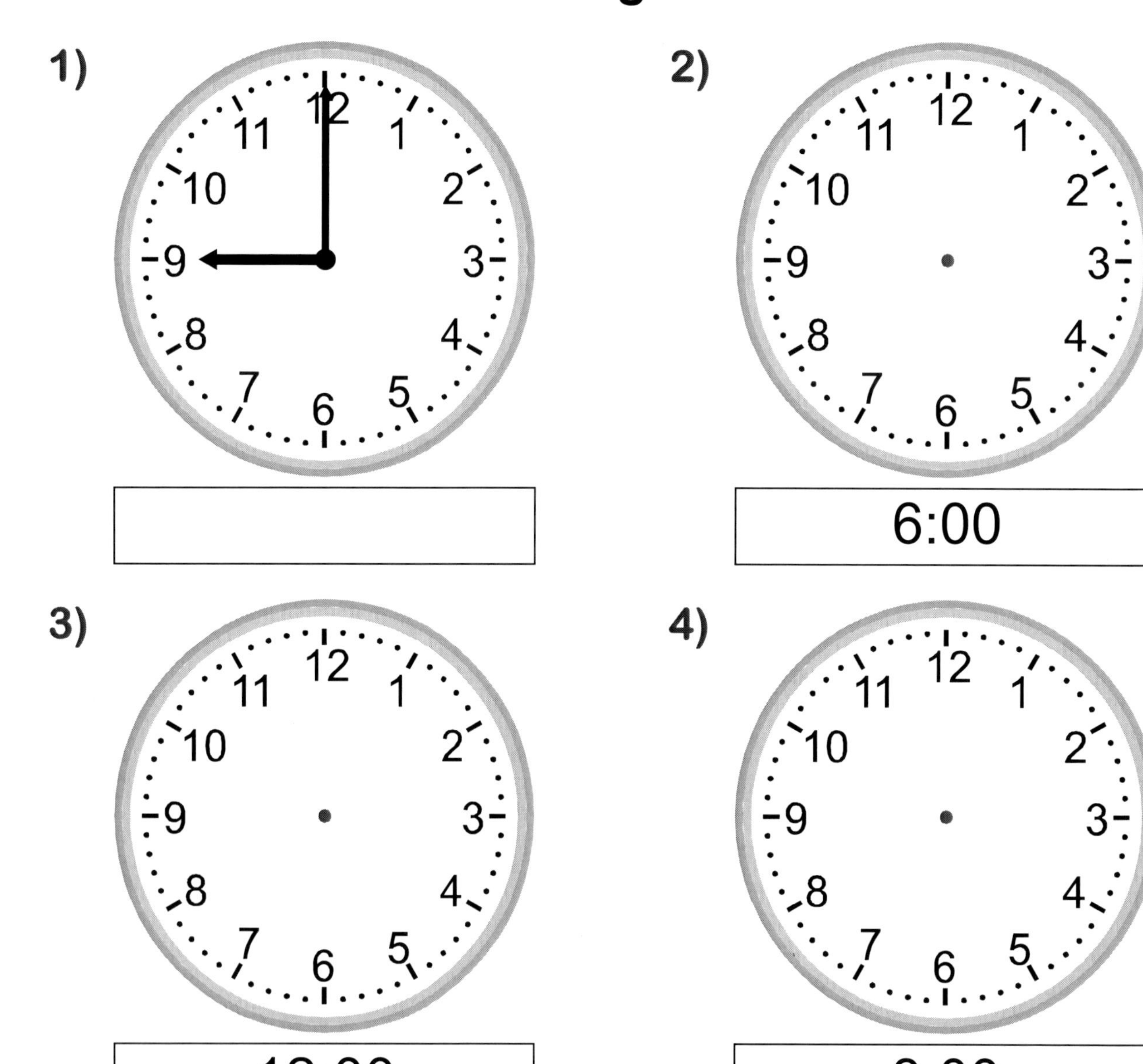

5)

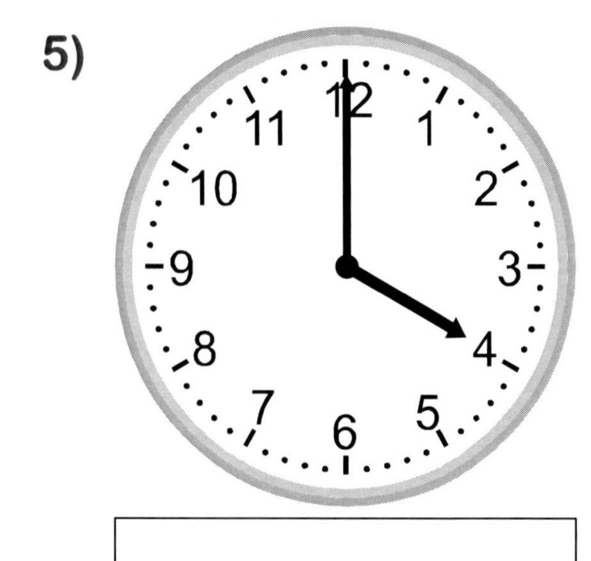

6)

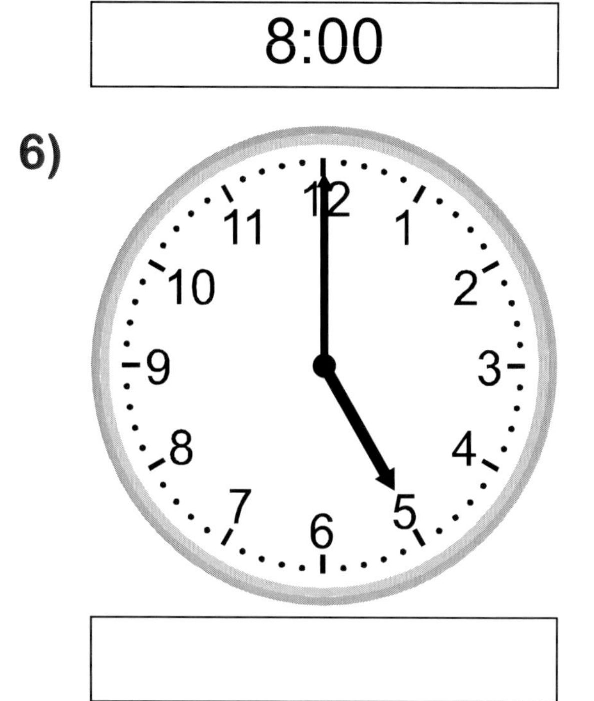

Page 35

1)

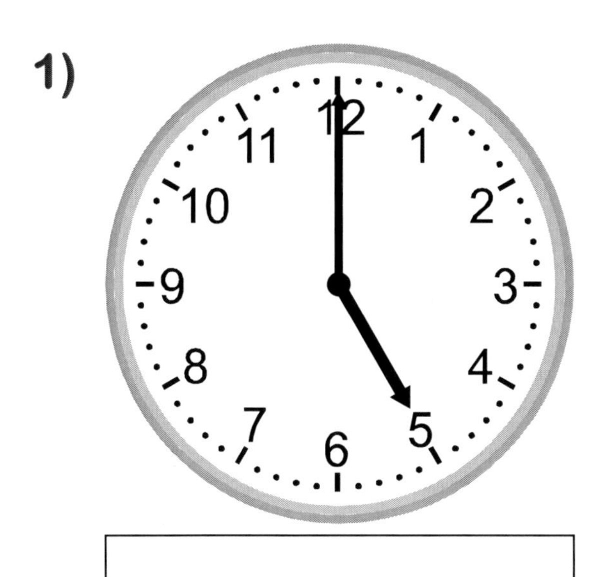

2)

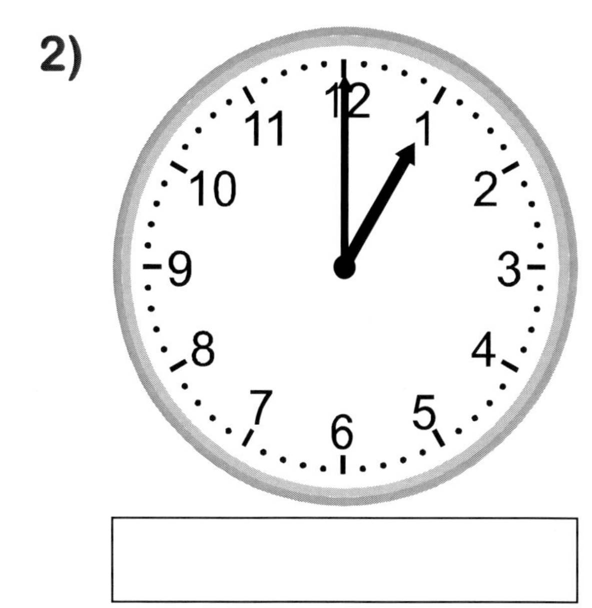

3)

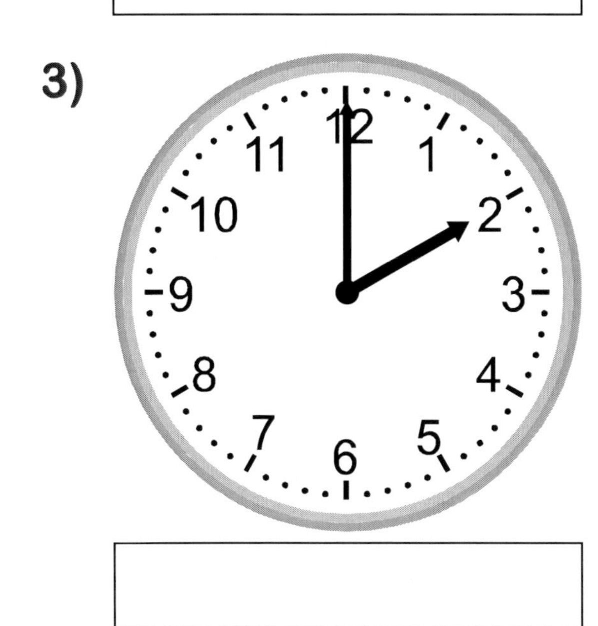

4)

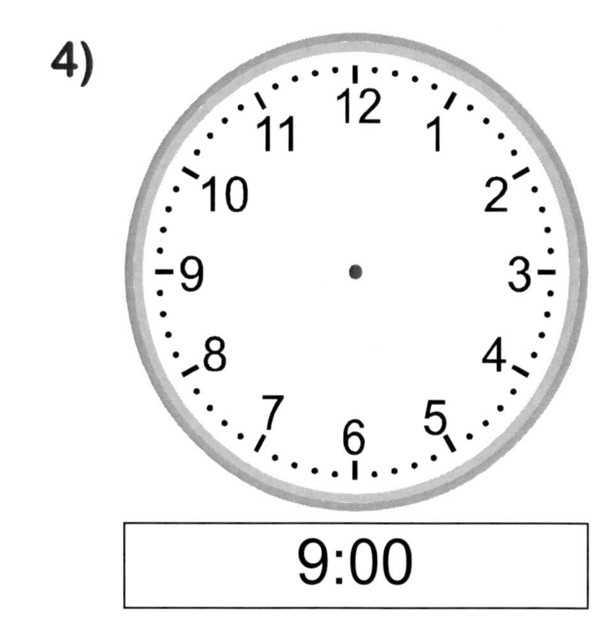

9:00

5)

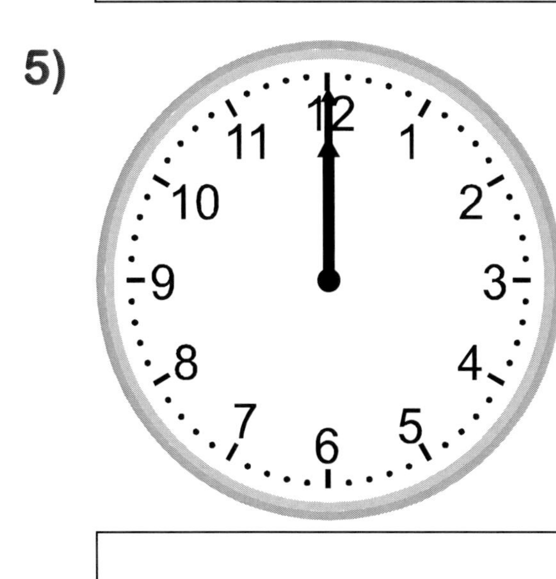

6)

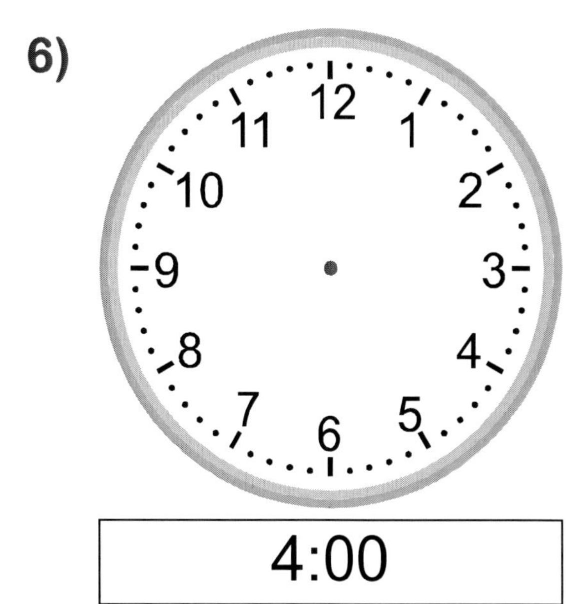

4:00

Page 36

1)

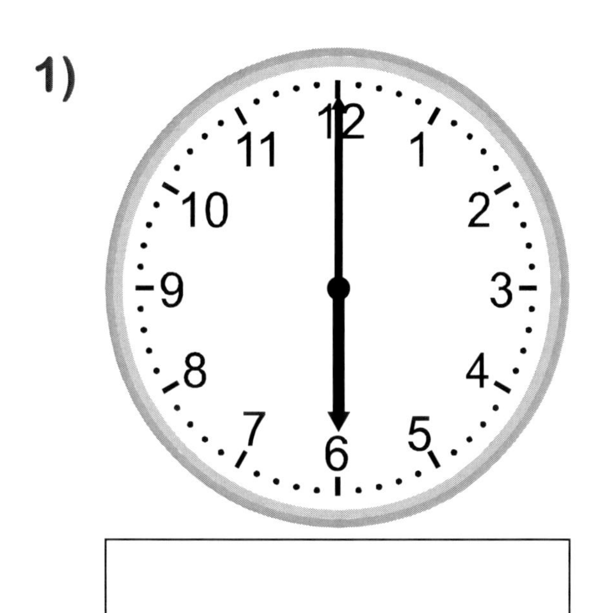

2)

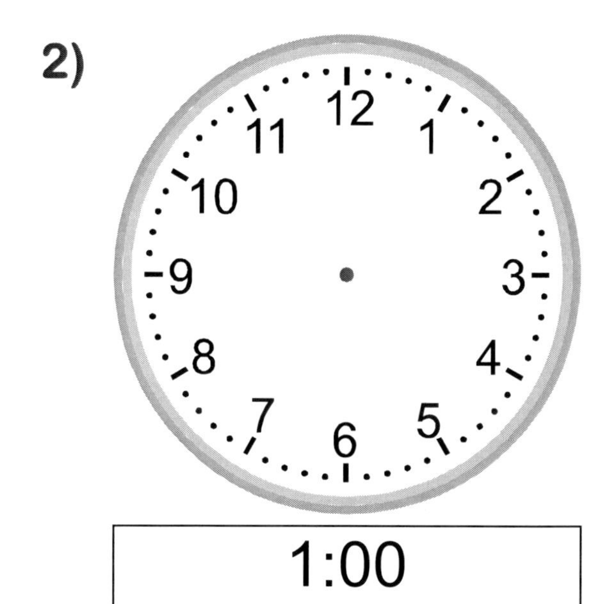

1:00

3)

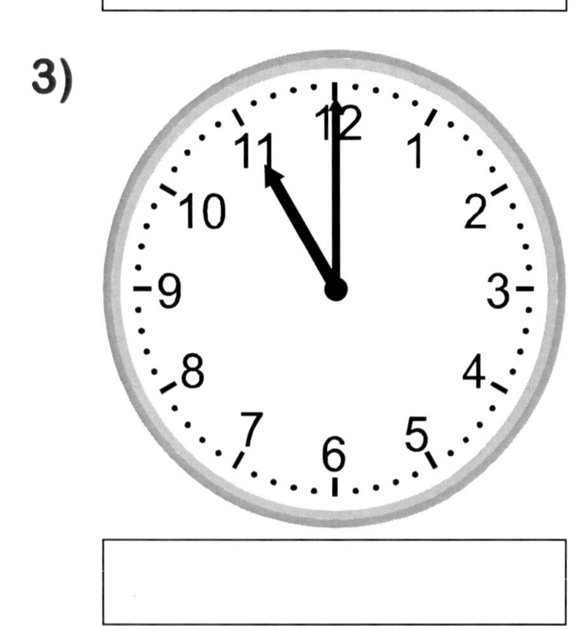

4)

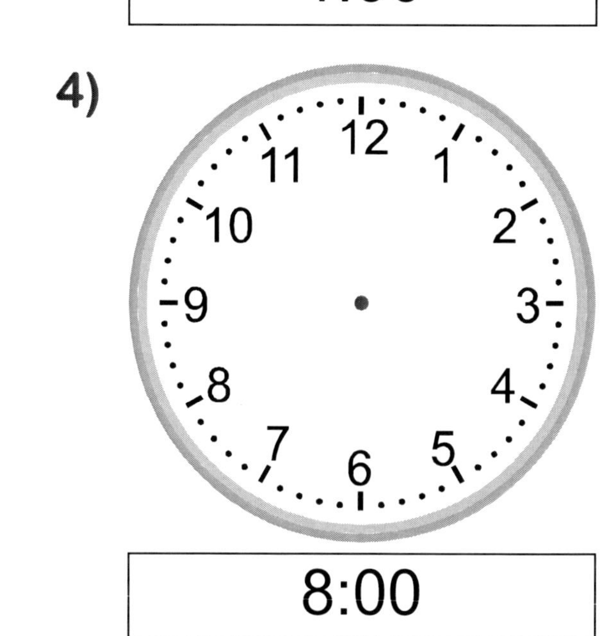

8:00

5)

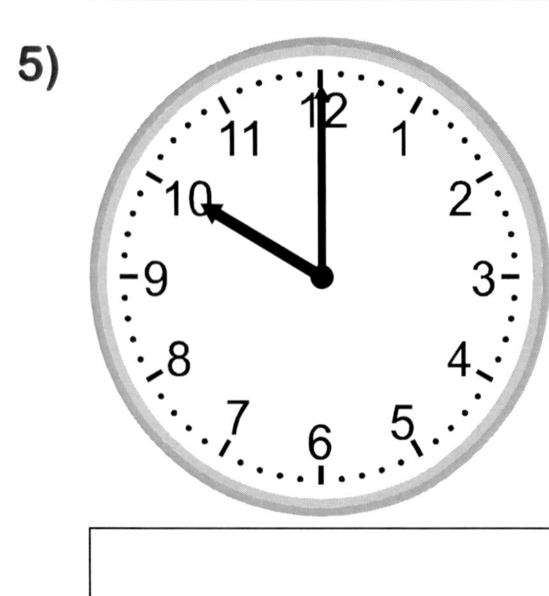

6)

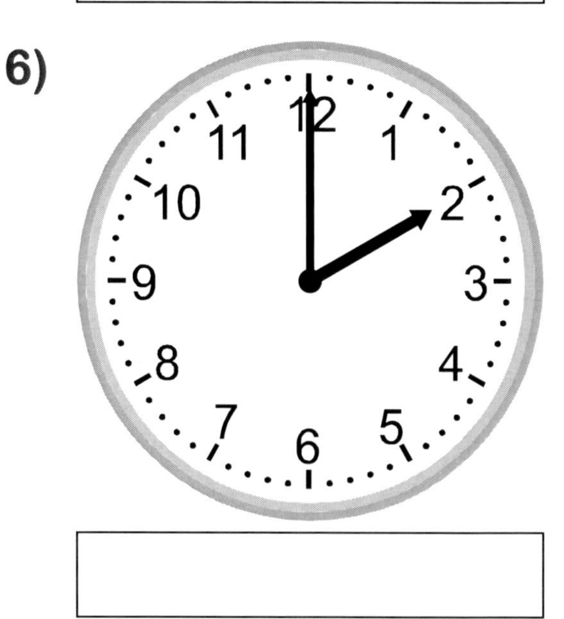

1)

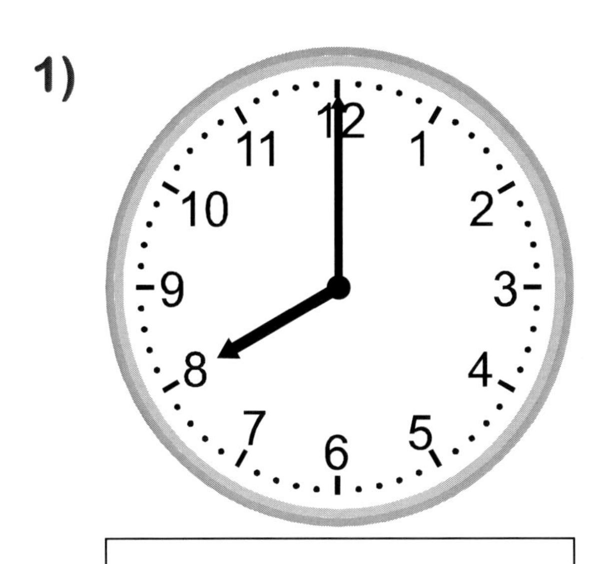

2)

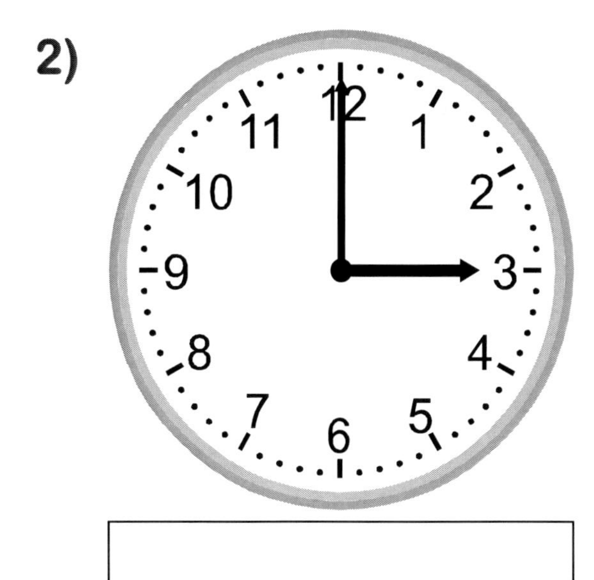

3)

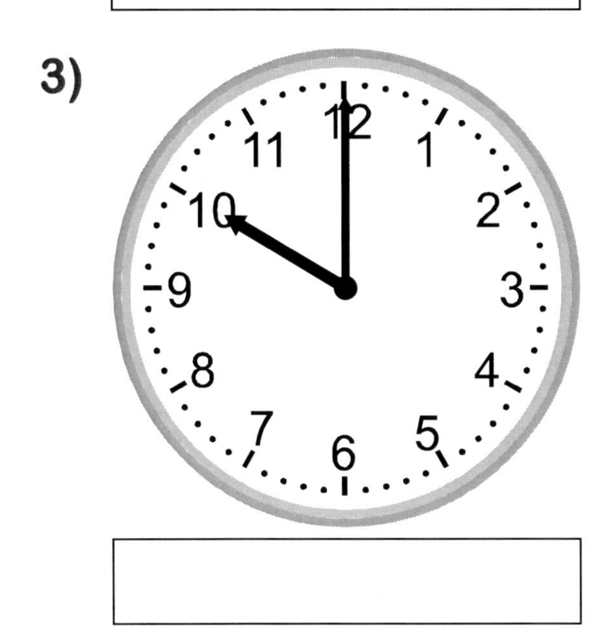

4)

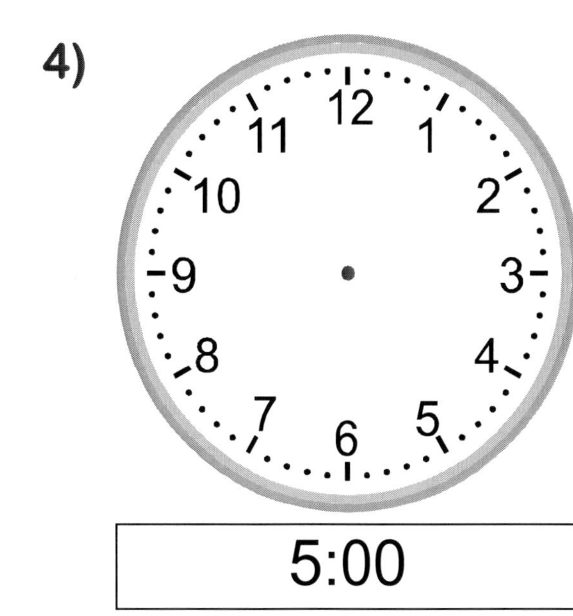

5)

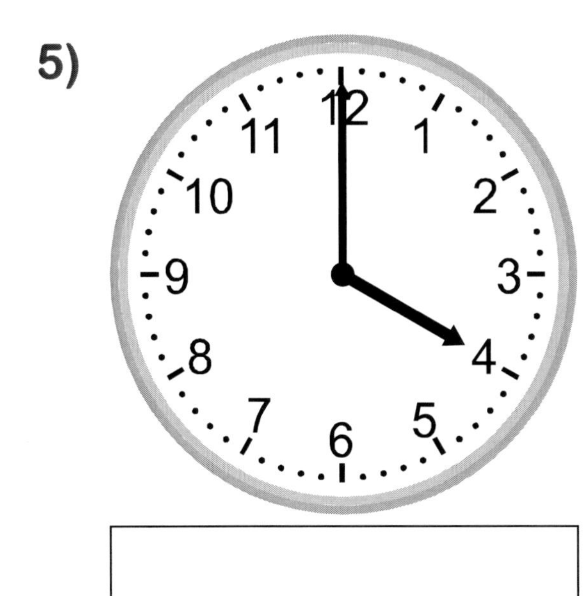

6)

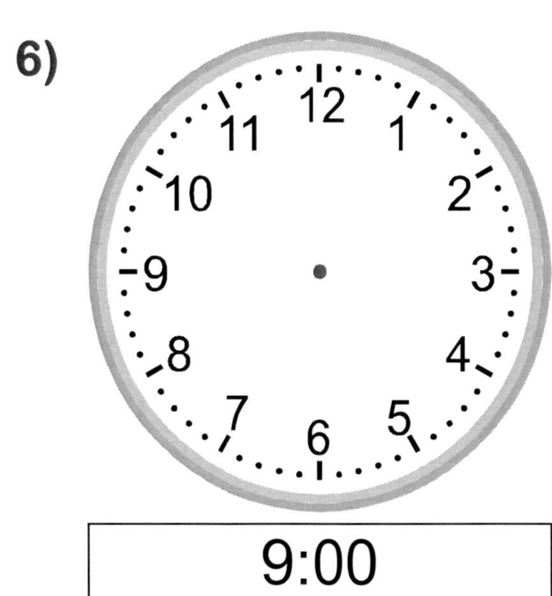

1)

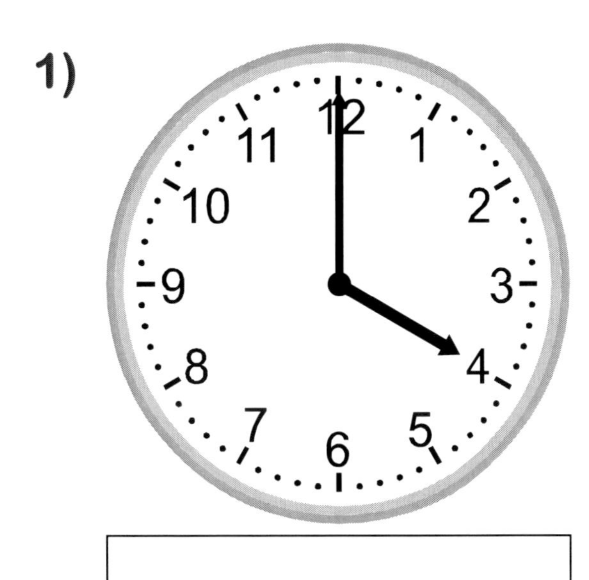

2)

3)

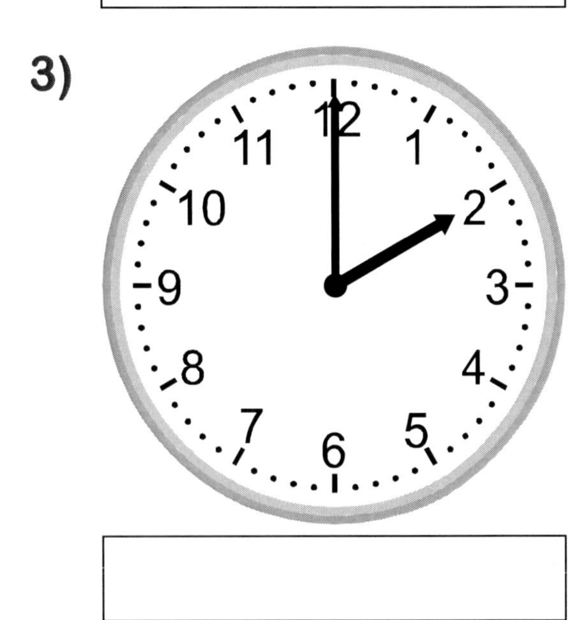

4)

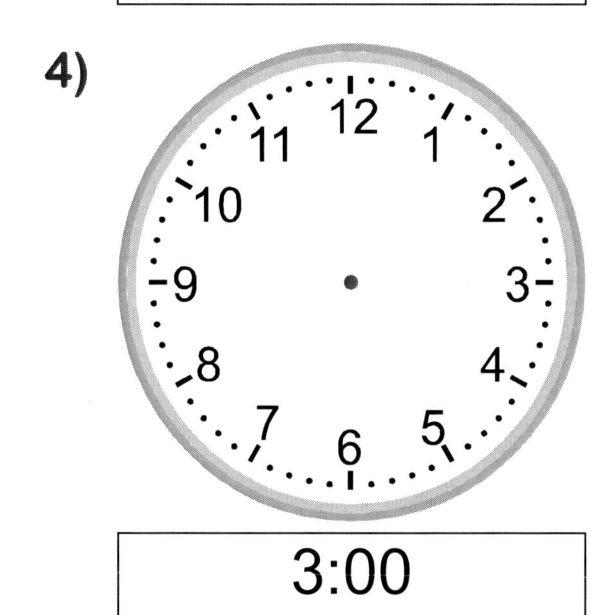

3:00

5)

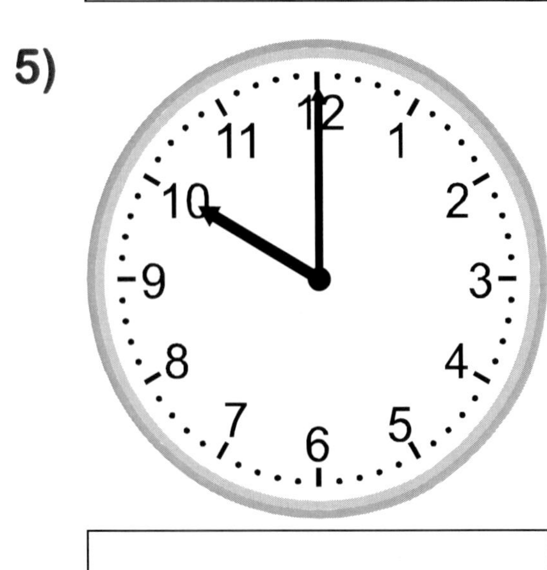

6)

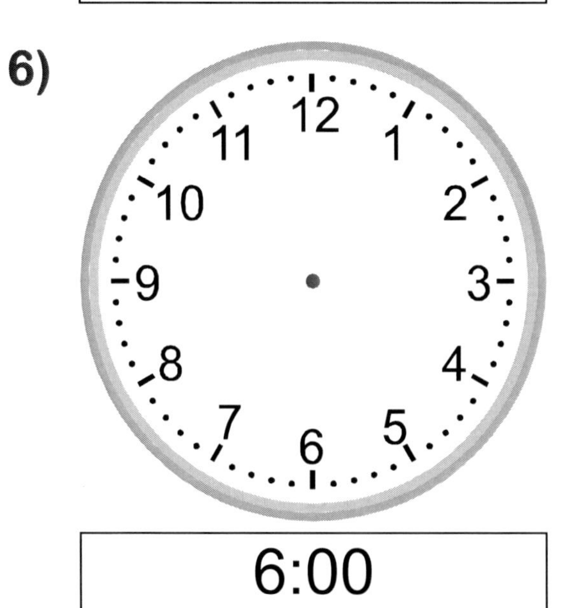

6:00

Page 39

1)

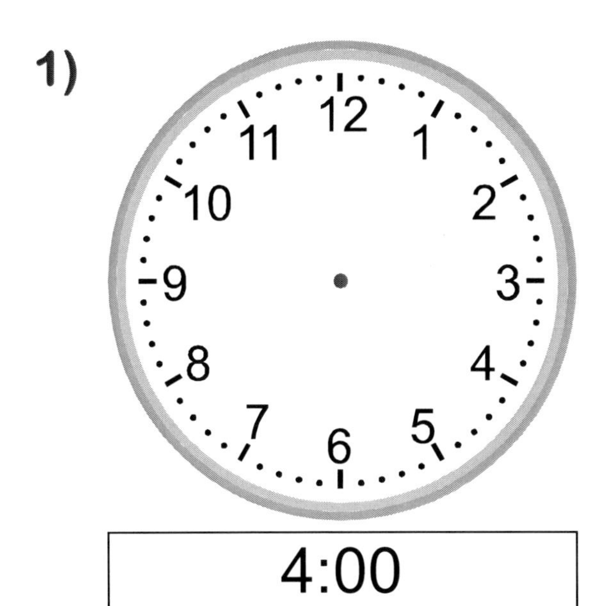

4:00

2)

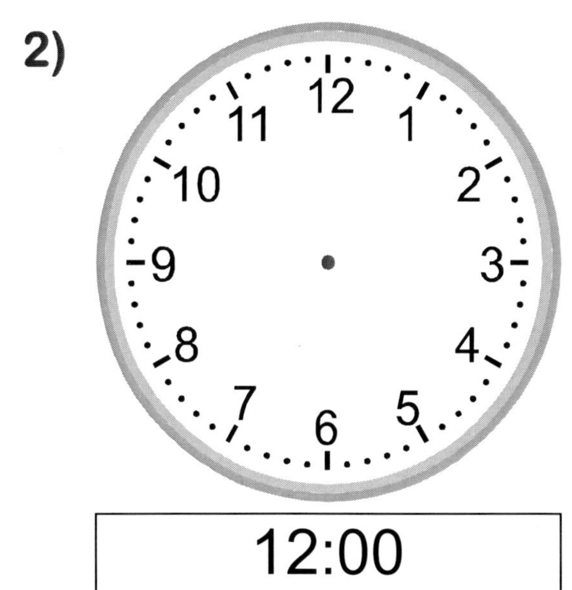

12:00

3)

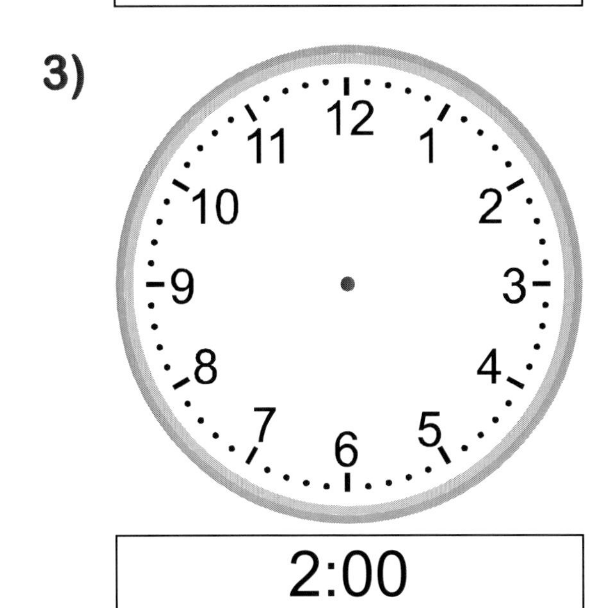

2:00

4)

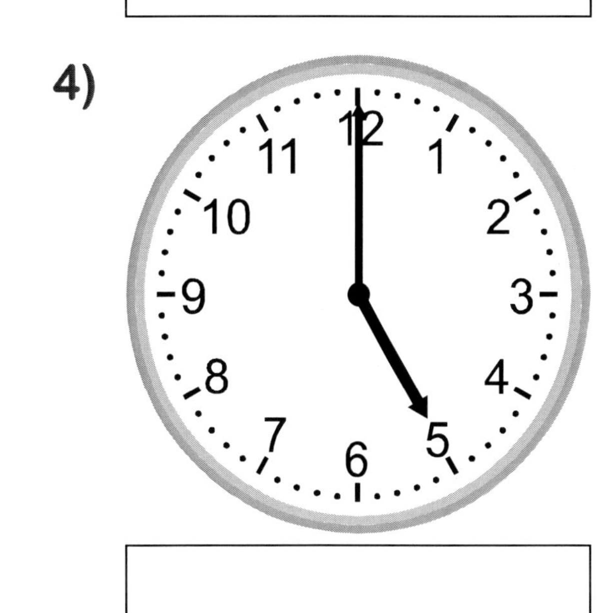

5)

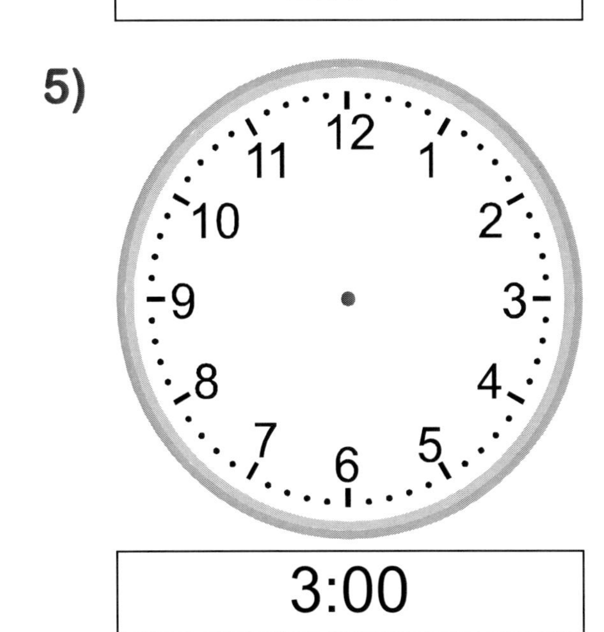

3:00

6)

1)

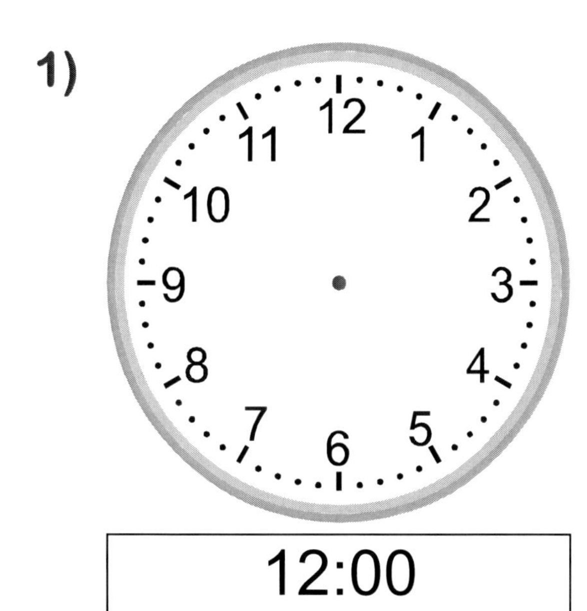

2)

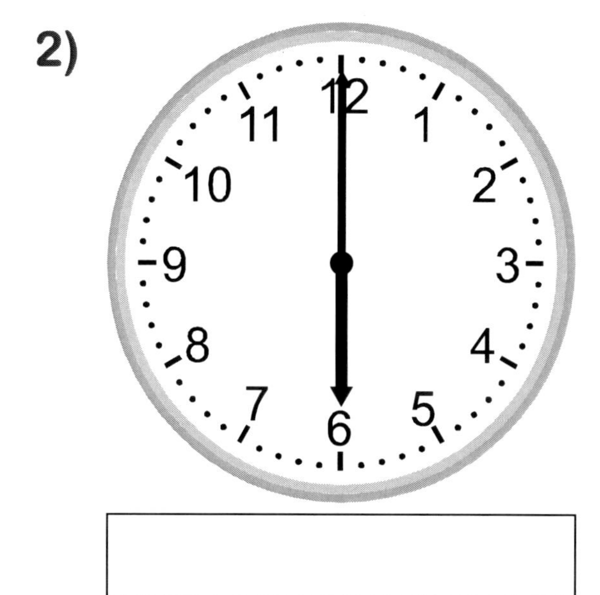

3)

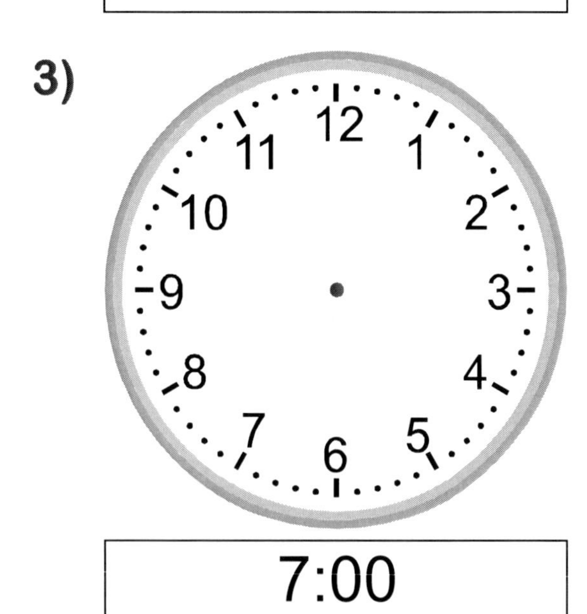

4)

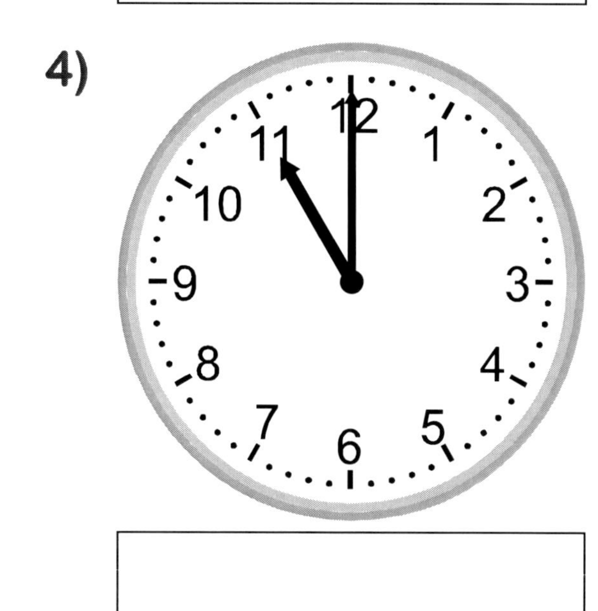

5)

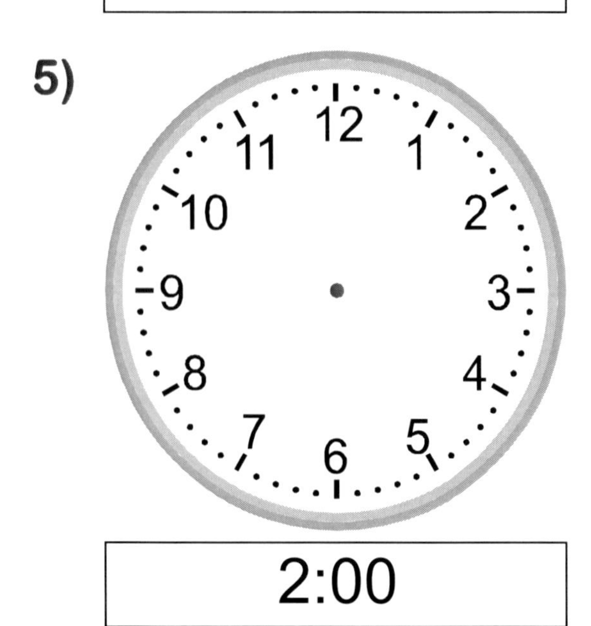

6)

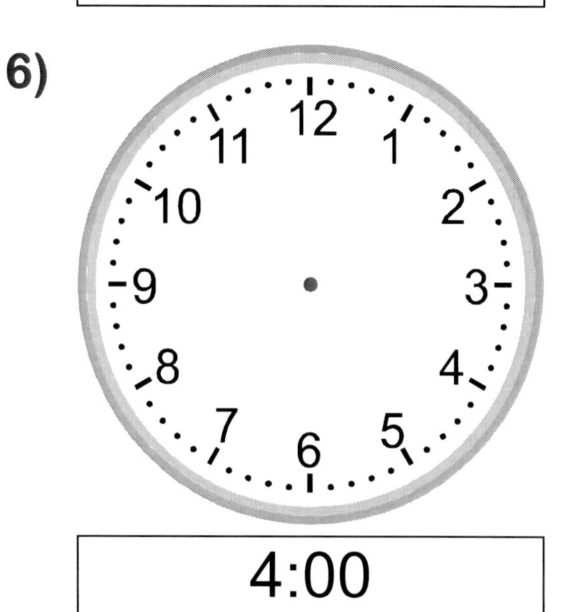

EXERCISES :

- COUNTING MONEY
- SHOPPING PROBLEMS

Page 41

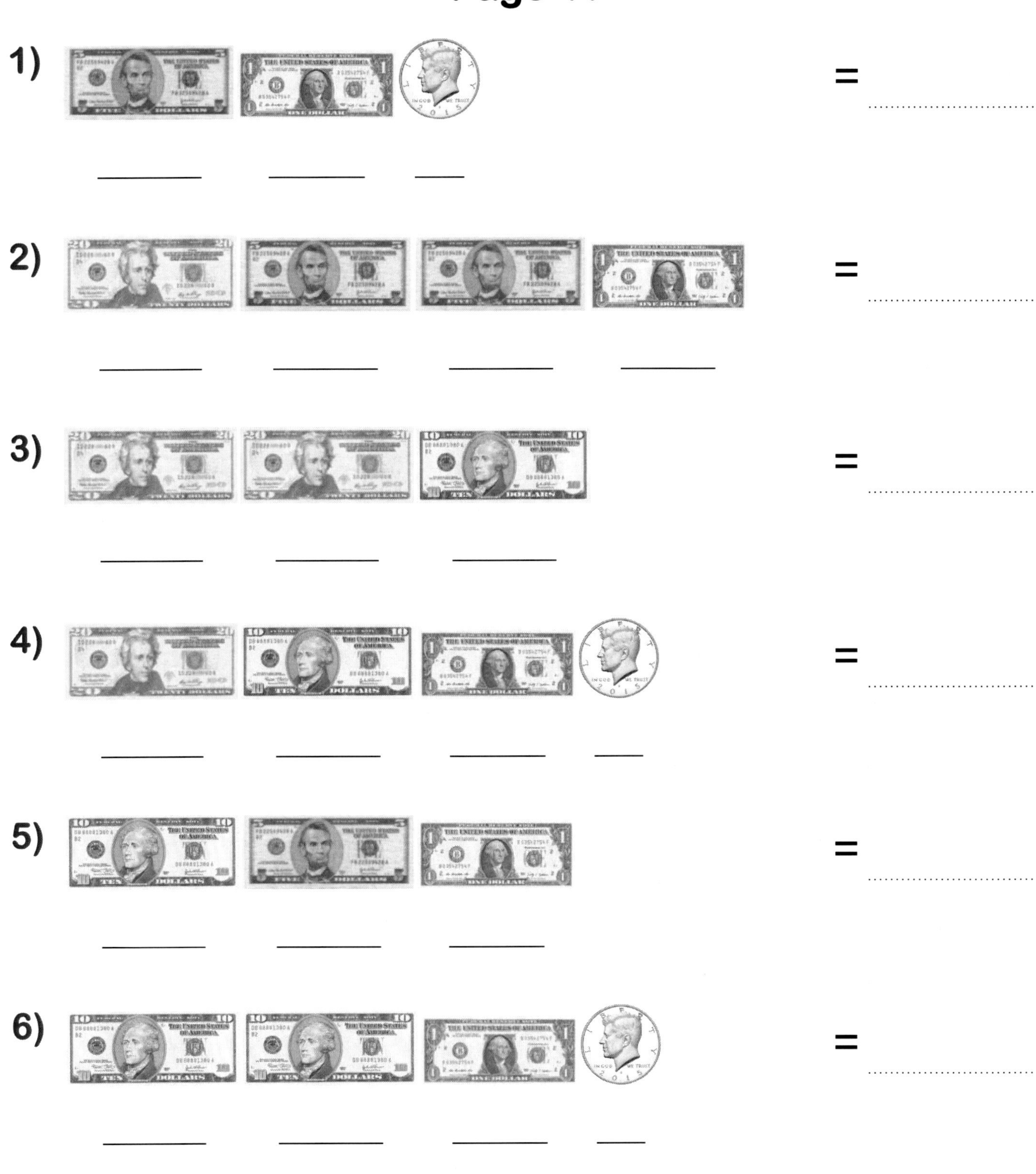

1) = ...

2) = ...

3) = ...

4) = ...

5) = ...

6) = ...

7) = ...

Page 42

1)  =

2)  =

3)  =

4)  =

5)  =

6)  =

7)  =

Page 43

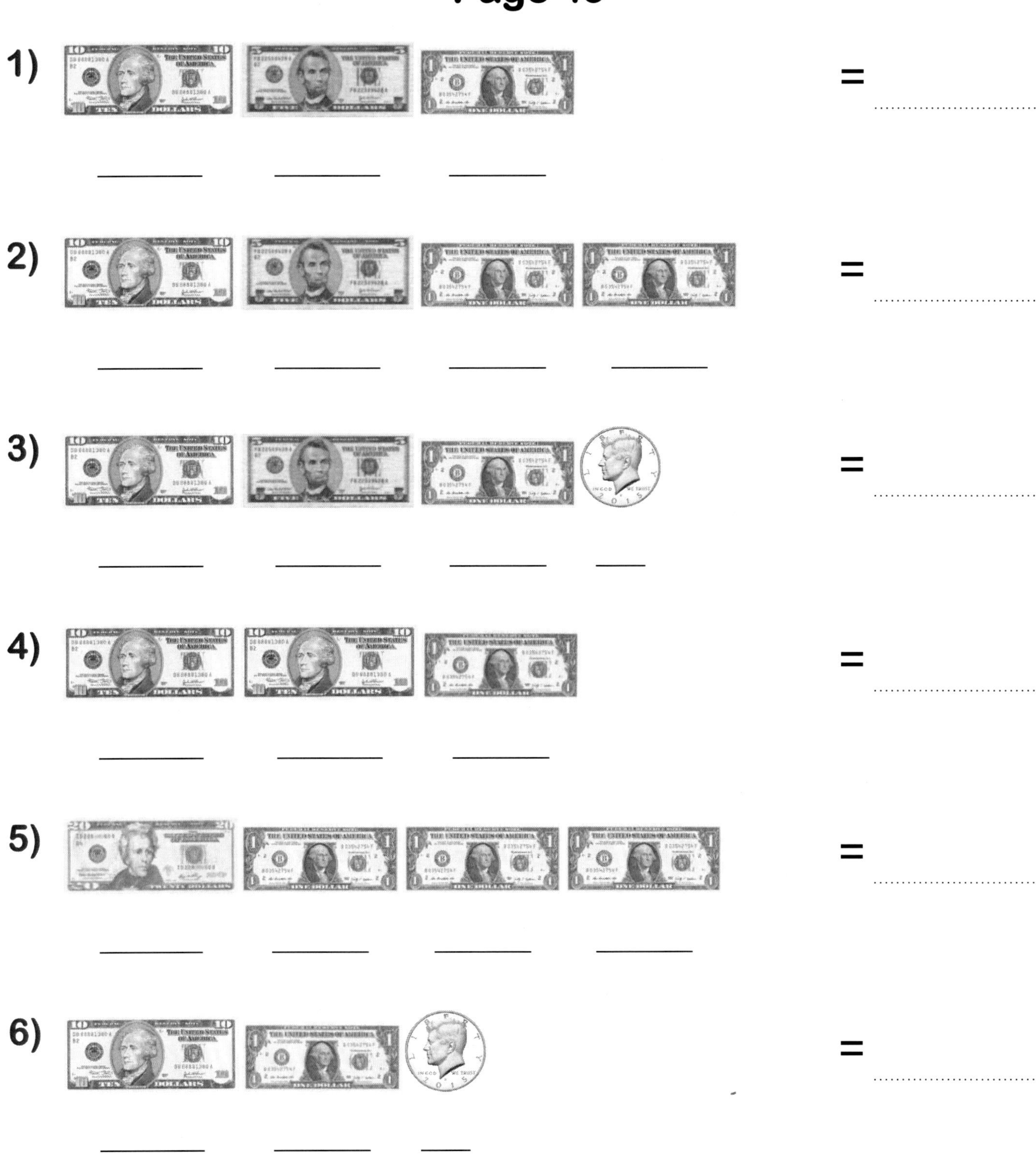

1) =

2) =

3) =

4) =

5) =

6) =

7) =

Page 44

1)  = ...

2)  = ...

3)  = ...

4)  = ...

5)  = ...

6)  = ...

7)  = ...

Page 45

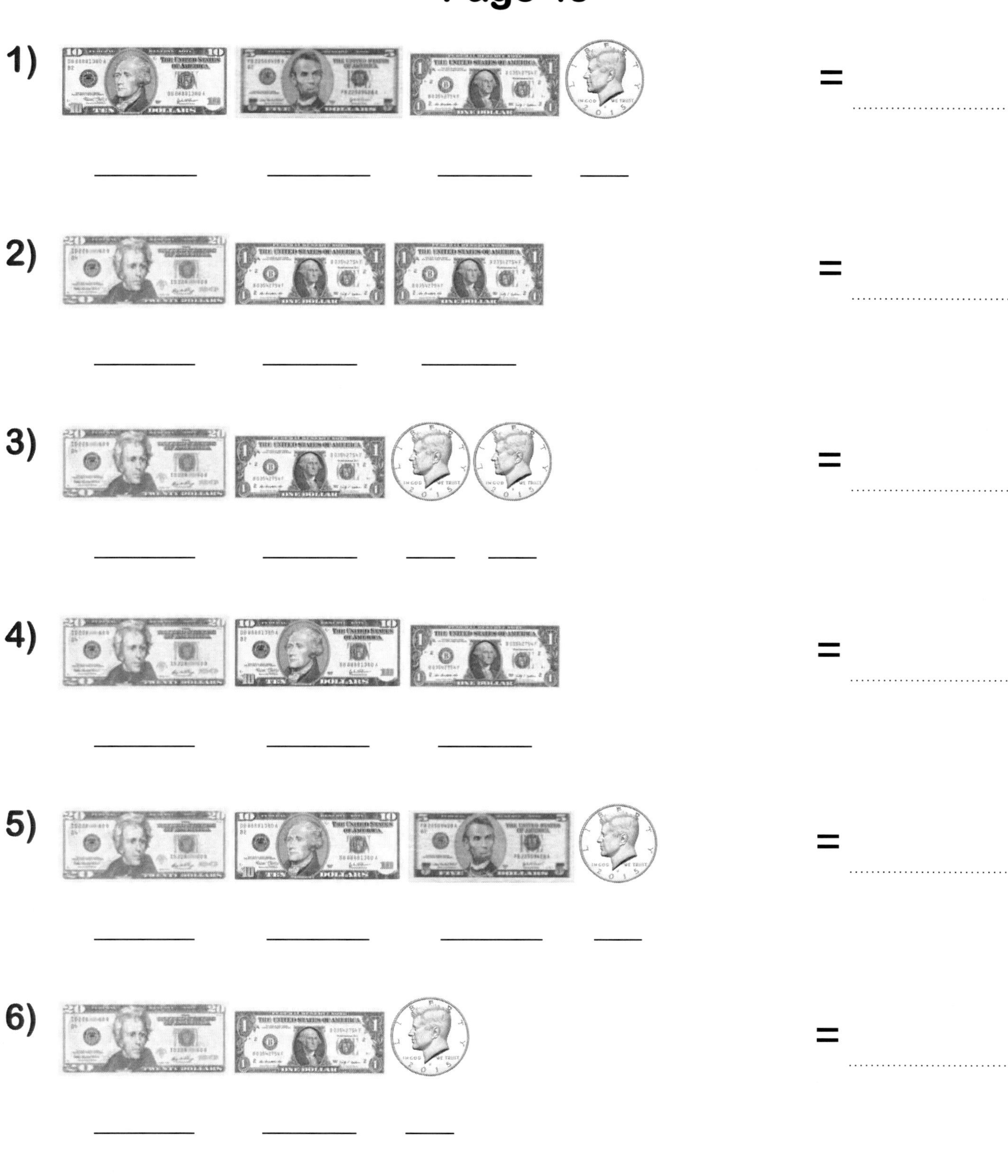

1) = ...

2) = ...

3) = ...

4) = ...

5) = ...

6) = ...

7) = ...

hot dog = \$1.00
order of French-fries = \$1.00
hamburger = \$2.00
deluxe cheeseburger = \$3.00
cola = \$1.00
ice cream cone = \$1.00
milk shake = \$2.00
taco = \$2.00

1) .... If Janet wanted to buy an ice cream cone and four tacos, how much money would she need?

2) .... Amy wants to buy a deluxe cheeseburger, three colas, and four hamburgers. How much will it cost her?

3) .... If Billy buys four ice cream cones and three orders of French-fries, how much change will he get back from \$10.00?

4) .... Sharon purchases five deluxe cheeseburgers, five hot dogs, and five orders of French-fries. If she had \$30.00, how much money will she have left?

hot dog = $1.00
order of French-fries = $1.00
hamburger = $2.00
deluxe cheeseburger = $3.00
cola = $1.00
ice cream cone = $1.00
milk shake = $2.00
taco = $2.00

1) .... Jake purchases four orders of French-fries and a cola. How much change will he get back from $10.00?

2) .... Paul wants to buy five tacos and four hamburgers. How much money will he need?

3) .... If Adam wanted to buy a milk shake, two colas, five ice cream cones, two hamburgers, and four tacos, how much money would he need?

4) .... Jennifer wants to buy two hot dogs, four tacos, five ice cream cones, two colas, and five deluxe cheeseburgers. How much will it cost her?

hot dog = $1.00
order of French-fries = $0.00
hamburger = $2.00
deluxe cheeseburger = $3.00
cola = $1.00
ice cream cone = $1.00
milk shake = $2.00
taco = $2.00

1) .... If Ellen wanted to buy five deluxe cheeseburgers, how much money would she need?

2) .... What is the total cost of four milk shakes?

3) .... If Jake wanted to buy four milk shakes, four colas, two orders of French-fries, three hot dogs, and three hamburgers, how much would he have to pay?

4) .... Brian wants to buy four ice cream cones, two orders of French-fries, four tacos, and two deluxe cheeseburgers. How much will he have to pay?

hot dog = $1.00
order of French-fries = $1.00
hamburger = $2.00
deluxe cheeseburger = $3.00
cola = $1.00
ice cream cone = $1.00
milk shake = $2.00
taco = $2.00

1) .... If Steven wanted to buy four hamburgers, a deluxe cheeseburger, three tacos, and four milk shakes, how much would he have to pay?

2) .... Ellen purchases four hamburgers and a cola. How much change will she get back from $15.00?

3) .... If Jackie wanted to buy four orders of French-fries, two deluxe cheeseburgers, four hamburgers, and four hot dogs, how much would she have to pay?

4) .... Sharon wants to buy three tacos, five milk shakes, and two colas. How much money will she need?

hot dog = $1.00
order of French-fries = $1.00
hamburger = $2.00
deluxe cheeseburger = $3.00
cola = $1.00
ice cream cone = $1.00
milk shake = $2.00
taco = $2.00

1) .... Adam wants to buy four deluxe cheeseburgers and two hamburgers. How much money will he need?

2) .... Jackie wants to buy two tacos. How much money will she need?

3) .... What is the total cost of a hamburger, three hot dogs, and four colas?

4) .... Janet purchases four hamburgers and four orders of French-fries. If she had $15.00, how much money will she have left?

FRACTIONS :

- COMPARING FRACTIONS
- FRACTION IDENTIFICATION

Page 51

1) $\frac{5}{8}$ .... $\frac{3}{8}$

2) $\frac{3}{4}$ .... $\frac{1}{4}$

3) $\frac{1}{5}$ .... $\frac{4}{5}$

4) $\frac{8}{6}$ .... $\frac{17}{6}$

5) $\frac{15}{21}$ .... $\frac{3}{21}$

6) $\frac{5}{9}$ .... $\frac{11}{9}$

7) $\frac{23}{10}$ .... $\frac{7}{10}$

8) $\frac{1}{2}$ .... $\frac{1}{2}$

9) $\frac{2}{6}$ .... $\frac{15}{6}$

10) $\frac{12}{8}$ .... $\frac{1}{8}$

11) $\frac{4}{3}$ .... $\frac{7}{3}$

12) $\frac{3}{4}$ .... $\frac{2}{4}$

13) $\frac{16}{7}$ .... $\frac{16}{7}$

14) $\frac{8}{24}$ .... $\frac{54}{24}$

15) $\frac{5}{8}$ .... $\frac{14}{8}$

16) $\frac{3}{2}$ .... $\frac{5}{2}$

17) $\frac{40}{45}$ .... $\frac{120}{45}$

18) $\frac{26}{10}$ .... $\frac{5}{10}$

19) $\frac{6}{5}$ .... $\frac{2}{5}$

20) $\frac{4}{7}$ .... $\frac{5}{7}$

21) $\frac{4}{10}$ .... $\frac{27}{10}$

Page 52

1) $\frac{5}{2}$ .... $\frac{1}{2}$

2) $\frac{15}{9}$ .... $\frac{8}{9}$

3) $\frac{2}{14}$ .... $\frac{12}{14}$

4) $\frac{15}{30}$ .... $\frac{13}{30}$

5) $\frac{3}{4}$ .... $\frac{1}{4}$

6) $\frac{2}{5}$ .... $\frac{1}{5}$

7) $\frac{14}{20}$ .... $\frac{49}{20}$

8) $\frac{6}{7}$ .... $\frac{13}{7}$

9) $\frac{24}{30}$ .... $\frac{76}{30}$

10) $\frac{8}{3}$ .... $\frac{2}{3}$

11) $\frac{6}{9}$ .... $\frac{3}{9}$

12) $\frac{3}{15}$ .... $\frac{44}{15}$

13) $\frac{3}{6}$ .... $\frac{4}{6}$

14) $\frac{7}{4}$ .... $\frac{1}{4}$

15) $\frac{11}{6}$ .... $\frac{1}{6}$

16) $\frac{2}{8}$ .... $\frac{7}{8}$

17) $\frac{4}{20}$ .... $\frac{11}{20}$

18) $\frac{1}{3}$ .... $\frac{7}{3}$

19) $\frac{15}{7}$ .... $\frac{13}{7}$

20) $\frac{1}{2}$ .... $\frac{1}{2}$

21) $\frac{3}{4}$ .... $\frac{3}{4}$

1) $\frac{1}{2}$ .... $\frac{3}{2}$

2) $\frac{10}{8}$ .... $\frac{14}{8}$

3) $\frac{1}{10}$ .... $\frac{3}{10}$

4) $\frac{6}{4}$ .... $\frac{2}{4}$

5) $\frac{3}{8}$ .... $\frac{22}{8}$

6) $\frac{10}{20}$ .... $\frac{34}{20}$

7) $\frac{20}{9}$ .... $\frac{11}{9}$

8) $\frac{6}{21}$ .... $\frac{5}{21}$

9) $\frac{8}{10}$ .... $\frac{7}{10}$

10) $\frac{21}{10}$ .... $\frac{1}{10}$

11) $\frac{23}{8}$ .... $\frac{7}{8}$

12) $\frac{2}{6}$ .... $\frac{10}{6}$

13) $\frac{1}{2}$ .... $\frac{1}{2}$

14) $\frac{7}{3}$ .... $\frac{5}{3}$

15) $\frac{10}{4}$ .... $\frac{6}{4}$

16) $\frac{21}{10}$ .... $\frac{9}{10}$

17) $\frac{15}{35}$ .... $\frac{27}{35}$

18) $\frac{6}{9}$ .... $\frac{6}{9}$

19) $\frac{6}{36}$ .... $\frac{80}{36}$

20) $\frac{2}{5}$ .... $\frac{7}{5}$

21) $\frac{3}{6}$ .... $\frac{3}{6}$

Page 54

1) $\frac{5}{8}$ .... $\frac{11}{8}$

2) $\frac{27}{10}$ .... $\frac{1}{10}$

3) $\frac{6}{5}$ .... $\frac{3}{5}$

4) $\frac{14}{9}$ .... $\frac{14}{9}$

5) $\frac{8}{28}$ .... $\frac{20}{28}$

6) $\frac{10}{6}$ .... $\frac{2}{6}$

7) $\frac{7}{10}$ .... $\frac{1}{10}$

8) $\frac{6}{8}$ .... $\frac{19}{8}$

9) $\frac{12}{48}$ .... $\frac{11}{48}$

10) $\frac{5}{2}$ .... $\frac{3}{2}$

11) $\frac{10}{15}$ .... $\frac{43}{15}$

12) $\frac{3}{2}$ .... $\frac{1}{2}$

13) $\frac{12}{5}$ .... $\frac{4}{5}$

14) $\frac{3}{7}$ .... $\frac{19}{7}$

15) $\frac{24}{36}$ .... $\frac{40}{36}$

16) $\frac{9}{30}$ .... $\frac{74}{30}$

17) $\frac{3}{4}$ .... $\frac{2}{4}$

18) $\frac{5}{3}$ .... $\frac{2}{3}$

19) $\frac{10}{40}$ .... $\frac{4}{40}$

20) $\frac{17}{6}$ .... $\frac{17}{6}$

21) $\frac{7}{8}$ .... $\frac{6}{8}$

1) $\frac{1}{5}$ .... $\frac{11}{5}$

2) $\frac{7}{3}$ .... $\frac{1}{3}$

3) $\frac{6}{12}$ .... $\frac{6}{12}$

4) $\frac{4}{8}$ .... $\frac{4}{8}$

5) $\frac{7}{4}$ .... $\frac{3}{4}$

6) $\frac{6}{14}$ .... $\frac{10}{14}$

7) $\frac{4}{18}$ .... $\frac{16}{18}$

8) $\frac{16}{10}$ .... $\frac{3}{10}$

9) $\frac{36}{48}$ .... $\frac{32}{48}$

10) $\frac{24}{9}$ .... $\frac{13}{9}$

11) $\frac{12}{30}$ .... $\frac{49}{30}$

12) $\frac{4}{3}$ .... $\frac{1}{3}$

13) $\frac{8}{10}$ .... $\frac{29}{10}$

14) $\frac{8}{6}$ .... $\frac{14}{6}$

15) $\frac{3}{4}$ .... $\frac{2}{4}$

16) $\frac{5}{7}$ .... $\frac{5}{7}$

17) $\frac{3}{2}$ .... $\frac{3}{2}$

18) $\frac{7}{3}$ .... $\frac{2}{3}$

19) $\frac{6}{7}$ .... $\frac{13}{7}$

20) $\frac{4}{5}$ .... $\frac{9}{5}$

21) $\frac{4}{8}$ .... $\frac{1}{8}$

1) $\frac{1}{3} =$ ☐☐☐

2) $\frac{4}{5} =$ ☐☐☐☐☐

3) $\frac{1}{2} =$ ☐☐

4) $\frac{2}{5} =$ ☐☐☐☐☐

5) $\frac{3}{4} =$ ☐☐☐☐

6) $\frac{1}{4} =$ ☐☐☐☐

7) $\frac{2}{3} =$ ☐☐☐

8) $\frac{3}{5} =$ ☐☐☐☐☐

9) $\frac{1}{5} =$ ☐☐☐☐☐

10) $\frac{2}{4} =$ ☐☐☐☐

11) $\frac{3}{5} =$ ☐☐☐☐☐

12) $\frac{3}{4} =$ ☐☐☐☐

13) $\frac{3}{5} =$ ☐☐☐☐☐

14) $\frac{2}{5} =$ ☐☐☐☐☐

1) $\frac{1}{3}$ = | | | |

2) $\frac{1}{2}$ = | | |

3) $\frac{2}{5}$ = | | | | | |

4) $\frac{3}{4}$ = | | | | |

5) $\frac{2}{3}$ = | | | |

6) $\frac{2}{4}$ = | | | | |

7) $\frac{1}{5}$ = | | | | | |

8) $\frac{3}{5}$ = | | | | | |

9) $\frac{1}{4}$ = | | | | |

10) $\frac{4}{5}$ = | | | | | |

11) $\frac{1}{3}$ = | | | |

12) $\frac{1}{5}$ = | | | | | |

13) $\frac{1}{4}$ = | | | | |

14) $\frac{1}{4}$ = | | | | |

Page 58

1) $\frac{2}{3} =$

2) $\frac{1}{2} =$

3) $\frac{2}{4} =$

4) $\frac{1}{5} =$

5) $\frac{1}{4} =$

6) $\frac{1}{3} =$

7) $\frac{3}{4} =$

8) $\frac{2}{5} =$

9) $\frac{3}{5} =$

10) $\frac{4}{5} =$

11) $\frac{2}{3} =$

12) $\frac{1}{2} =$

13) $\frac{1}{2} =$

14) $\frac{3}{5} =$

1) $\frac{3}{4}$ =

2) $\frac{1}{5}$ =

3) $\frac{1}{3}$ =

4) $\frac{1}{2}$ =

5) $\frac{3}{5}$ =

6) $\frac{2}{3}$ =

7) $\frac{2}{5}$ =

8) $\frac{2}{4}$ =

9) $\frac{4}{5}$ =

10) $\frac{1}{4}$ =

11) $\frac{2}{3}$ =

12) $\frac{4}{5}$ =

13) $\frac{3}{5}$ =

14) $\frac{1}{3}$ =

Page 60

1) $\frac{1}{5}$ =

2) $\frac{2}{5}$ =

3) $\frac{1}{2}$ =

4) $\frac{1}{3}$ =

5) $\frac{3}{4}$ =

6) $\frac{2}{4}$ =

7) $\frac{2}{3}$ =

8) $\frac{4}{5}$ =

9) $\frac{3}{5}$ =

10) $\frac{1}{4}$ =

11) $\frac{4}{5}$ =

12) $\frac{2}{5}$ =

13) $\frac{1}{2}$ =

14) $\frac{1}{2}$ =

MEASUREMENT :

- METRIC CONVERSION
- US WEIGHTS AND MESURES

1) 8 ft = ... m

2) 13 ft = ... m

3) 18 ft = ... m

4) 12 ft = ... m

5) 16 ft = ... m

6) 12 ft = ... m

7) 13 ft = ... m

8) 9 ft = ... m

9) 10 ft = ... m

10) 9 ft = ... m

11) 16 ft = ... m

12) 8 ft = ... m

13) 2 ft = ... m

14) 1 ft = ... m

15) 7 ft = ... m

16) 4 ft = ... m

17) 6 ft = ... m

18) 16 ft = ... m

19) 19 ft = ... m

20) 5 ft = ... m

1) 16 ft = ... m
2) 8 ft = ... m
3) 19 ft = ... m
4) 15 ft = ... m
5) 6 ft = ... m
6) 3 ft = ... m
7) 13 ft = ... m
8) 4 ft = ... m
9) 5 ft = ... m
10) 9 ft = ... m
11) 18 ft = ... m
12) 12 ft = ... m
13) 14 ft = ... m
14) 9 ft = ... m
15) 17 ft = ... m
16) 14 ft = ... m
17) 6 ft = ... m
18) 6 ft = ... m
19) 8 ft = ... m
20) 4 ft = ... m

1) 14 ft = ... m

2) 3 ft = ... m

3) 3 ft = ... m

4) 6 ft = ... m

5) 9 ft = ... m

6) 3 ft = ... m

7) 20 ft = ... m

8) 17 ft = ... m

9) 12 ft = ... m

10) 1 ft = ... m

11) 7 ft = ... m

12) 7 ft = ... m

13) 12 ft = ... m

14) 6 ft = ... m

15) 13 ft = ... m

16) 9 ft = ... m

17) 9 ft = ... m

18) 7 ft = ... m

19) 15 ft = ... m

20) 5 ft = ... m

1) 11 ft = ... m

2) 3 ft = ... m

3) 12 ft = ... m

4) 11 ft = ... m

5) 8 ft = ... m

6) 8 ft = ... m

7) 12 ft = ... m

8) 5 ft = ... m

9) 5 ft = ... m

10) 6 ft = ... m

11) 11 ft = ... m

12) 20 ft = ... m

13) 19 ft = ... m

14) 14 ft = ... m

15) 11 ft = ... m

16) 14 ft = ... m

17) 7 ft = ... m

18) 14 ft = ... m

19) 7 ft = ... m

20) 4 ft = ... m

1) 3 ft = ... m
2) 16 ft = ... m
3) 12 ft = ... m
4) 4 ft = ... m
5) 3 ft = ... m
6) 12 ft = ... m
7) 1 ft = ... m
8) 9 ft = ... m
9) 17 ft = ... m
10) 10 ft = ... m
11) 6 ft = ... m
12) 19 ft = ... m
13) 11 ft = ... m
14) 4 ft = ... m
15) 13 ft = ... m
16) 12 ft = ... m
17) 10 ft = ... m
18) 12 ft = ... m
19) 18 ft = ... m
20) 2 ft = ... m

1) 6 in = ... ft

2) 14 ft = ... yd

3) 8 in = ... yd

4) 8 ft = ... yd

5) 17 in = ... ft

6) 16 in = ... ft

7) 20 ft = ... yd

8) 11 in = ... yd

9) 5 ft = ... yd

10) 7 in = ... ft

11) 15 ft = ... yd

12) 1 in = ... yd

13) 5 in = ... yd

14) 18 in = ... ft

15) 13 in = ... ft

16) 11 ft = ... yd

17) 17 ft = ... yd

18) 4 in = ... yd

19) 19 ft = ... yd

20) 13 in = ... yd

1) 5 ft = ... yd

2) 9 ft = ... yd

3) 8 in = ... ft

4) 8 ft = ... yd

5) 16 ft = ... yd

6) 3 ft = ... yd

7) 15 ft = ... yd

8) 14 ft = ... yd

9) 11 ft = ... yd

10) 2 ft = ... yd

11) 14 in = ... yd

12) 19 in = ... ft

13) 15 in = ... ft

14) 4 in = ... ft

15) 15 in = ... yd

16) 12 in = ... ft

17) 6 in = ... yd

18) 13 in = ... yd

19) 6 ft = ... yd

20) 19 ft = ... yd

1) 6 in = ... ft

2) 17 ft = ... yd

3) 16 in = ... ft

4) 4 ft = ... yd

5) 6 in = ... yd

6) 18 ft = ... yd

7) 2 in = ... yd

8) 6 ft = ... yd

9) 1 ft = ... yd

10) 19 ft = ... yd

11) 2 ft = ... yd

12) 14 ft = ... yd

13) 15 ft = ... yd

14) 8 in = ... ft

15) 7 in = ... ft

16) 4 in = ... ft

17) 9 ft = ... yd

18) 12 in = ... yd

19) 12 in = ... ft

20) 17 in = ... ft

1) 8 ft = ... yd
2) 12 ft = ... yd
3) 4 in = ... yd
4) 14 ft = ... yd
5) 13 in = ... yd
6) 14 in = ... ft
7) 16 ft = ... yd
8) 9 ft = ... yd
9) 5 in = ... yd
10) 10 in = ... ft
11) 2 ft = ... yd
12) 18 in = ... yd
13) 15 in = ... ft
14) 15 ft = ... yd
15) 11 ft = ... yd
16) 6 in = ... yd
17) 19 in = ... ft
18) 20 in = ... ft
19) 7 ft = ... yd
20) 10 in = ... yd

1) 17 ft = ... yd
2) 12 ft = ... yd
3) 10 ft = ... yd
4) 10 in = ... ft
5) 14 ft = ... yd
6) 5 in = ... ft
7) 13 ft = ... yd
8) 9 in = ... yd
9) 11 ft = ... yd
10) 4 ft = ... yd
11) 7 ft = ... yd
12) 7 in = ... yd
13) 8 ft = ... yd
14) 6 ft = ... yd
15) 10 in = ... yd
16) 2 in = ... ft
17) 2 in = ... yd
18) 4 in = ... ft
19) 17 in = ... ft
20) 5 in = ... yd

GEOMETRY EXERCISES:
- PERIMETER AND AREA

Page 71

1)

22 ft

19 ft

....

2)

17 ft

10 ft

....

3)

22 ft

20 ft

....

4)

19 ft

20 ft

....

5)

28 ft

25 ft

....

6)

24 ft

19 ft

....

7)

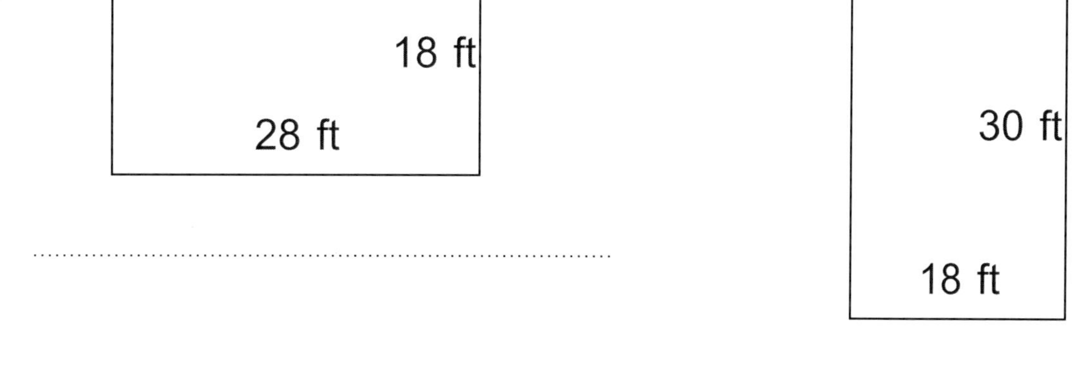

....

8)

....

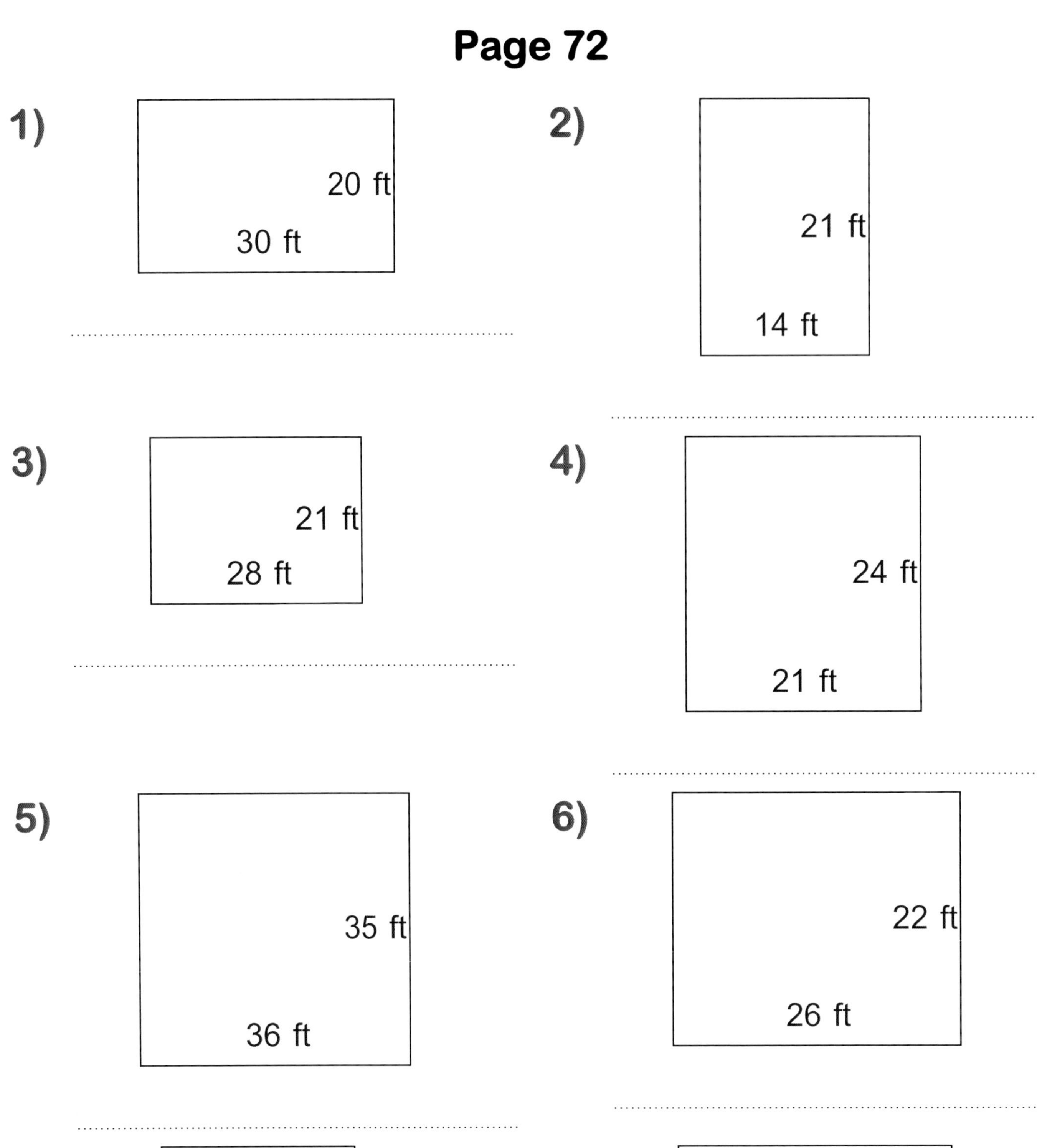
1)
20 ft
30 ft
2)
21 ft
14 ft
3)
21 ft
28 ft
4)
24 ft
21 ft
5)
35 ft
36 ft
6)
22 ft
26 ft

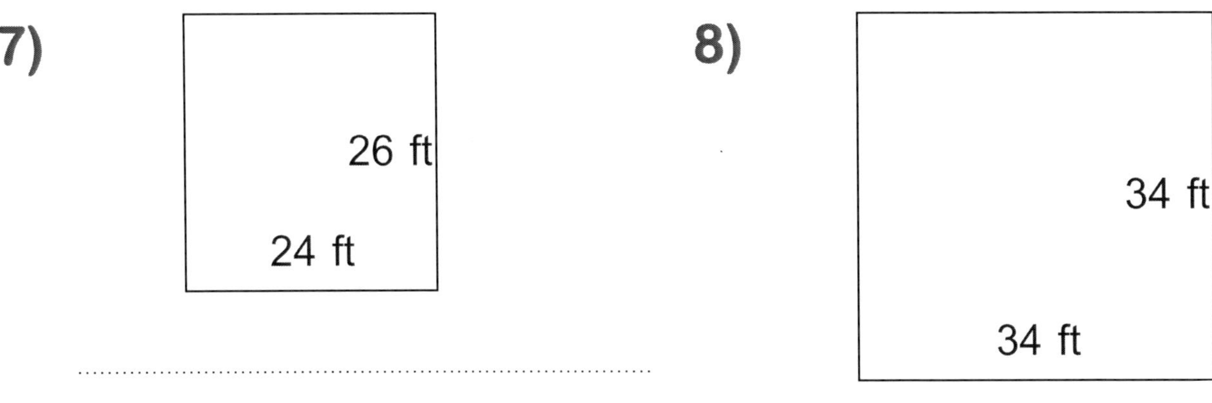
7)
26 ft
24 ft
8)
34 ft
34 ft

Page 73

1) 24 ft, 18 ft

....

2) 27 ft, 19 ft

....

3) 28 ft, 24 ft

....

4) 20 ft, 24 ft

....

5) 13 ft, 15 ft

....

6) 31 ft, 26 ft

....

7) 19 ft, 11 ft

....

8) 20 ft, 23 ft

....

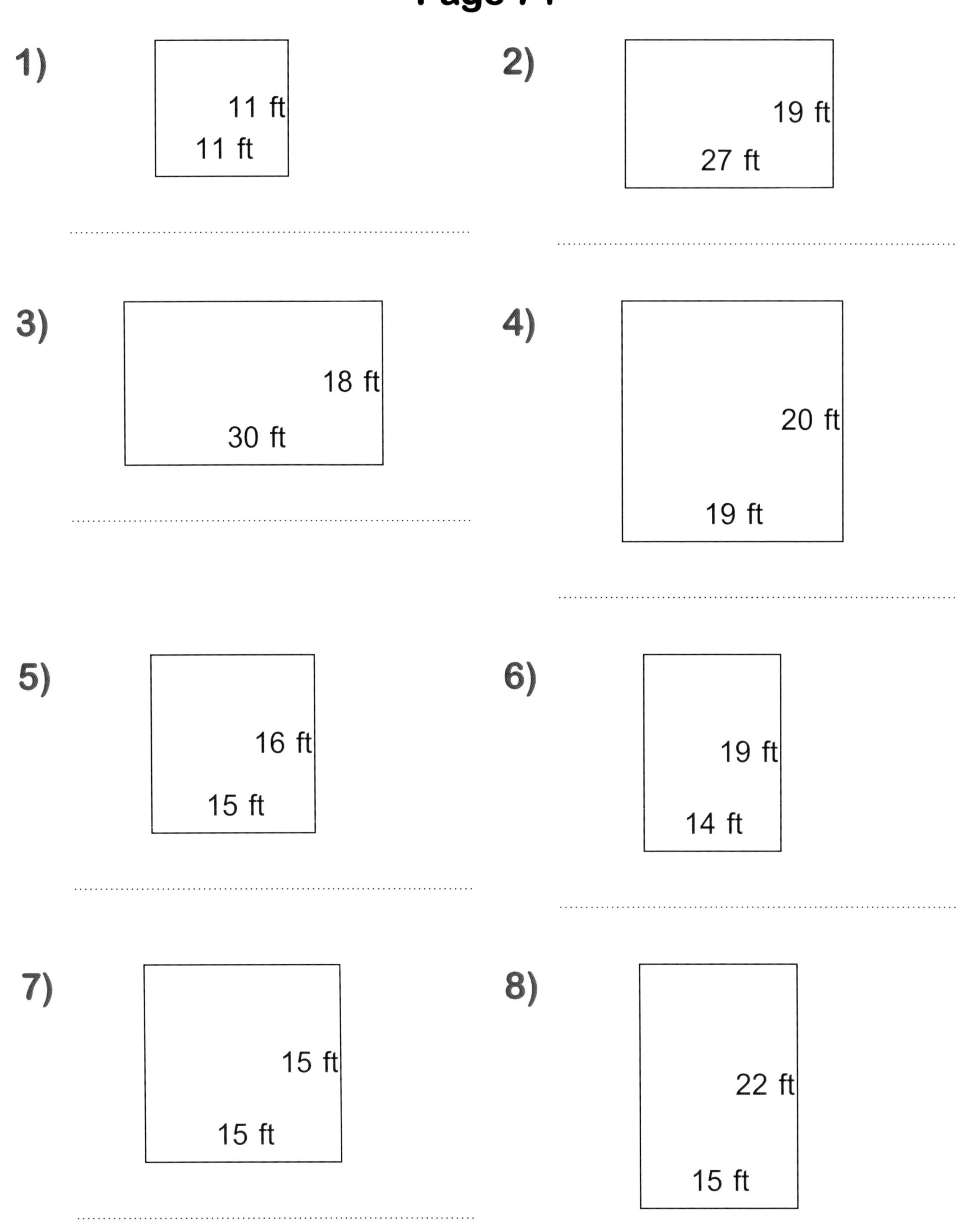
1)
11 ft
11 ft
2)
19 ft
27 ft
3)
18 ft
30 ft
4)
20 ft
19 ft
5)
16 ft
15 ft
6)
19 ft
14 ft
7)
15 ft
15 ft
8)
22 ft
15 ft

Page 75

1)

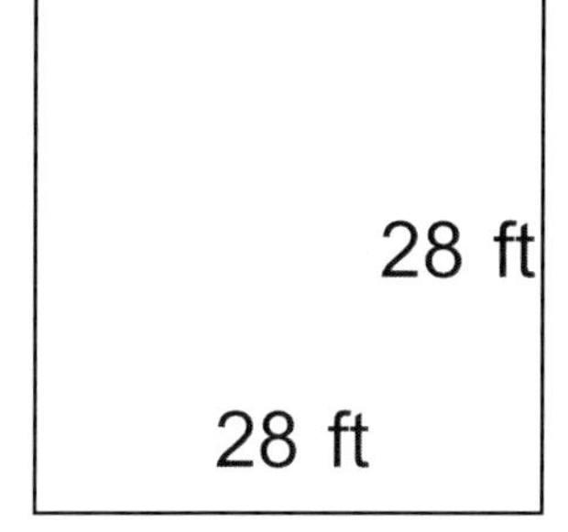

2)

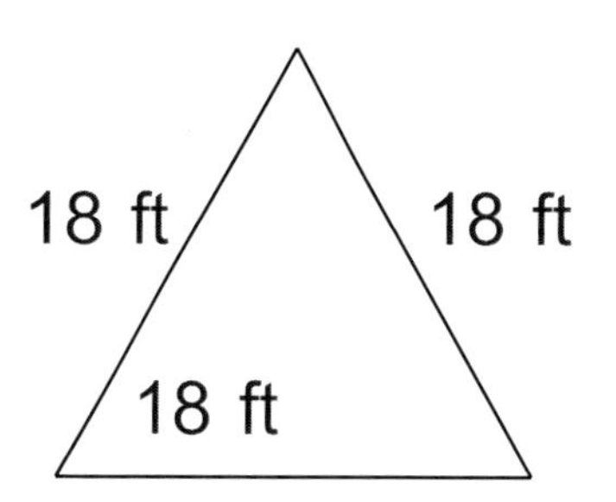

3)

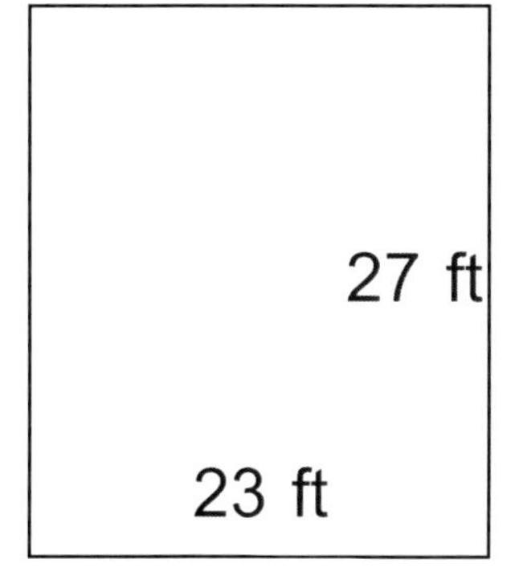

4)

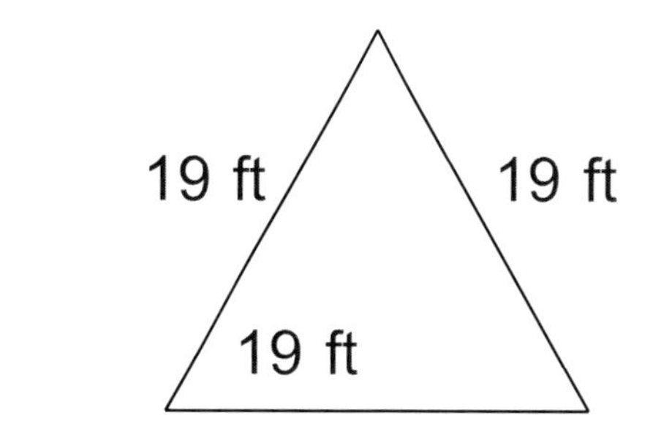

5)

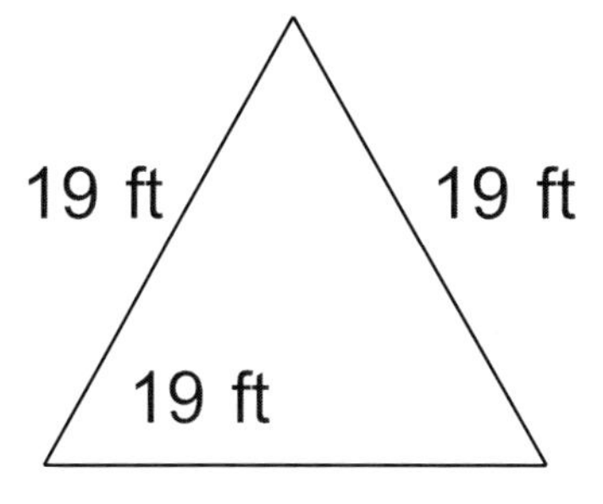

6)

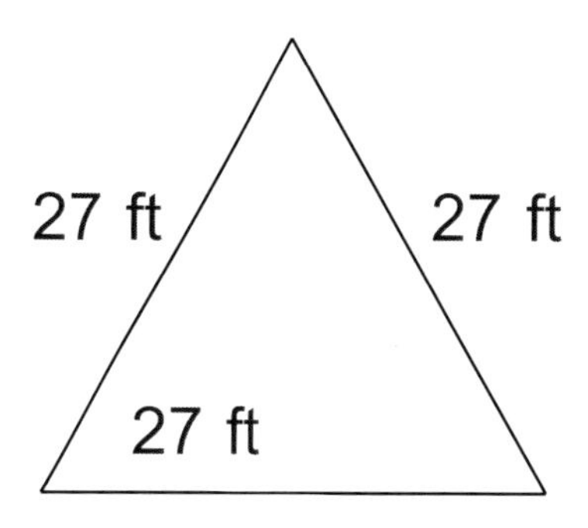

7)

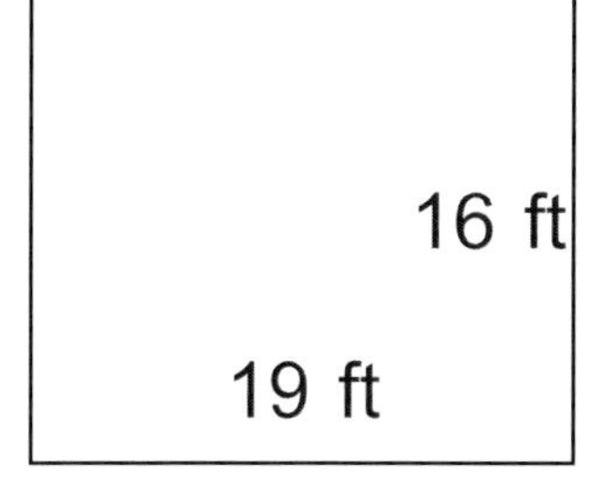

8)

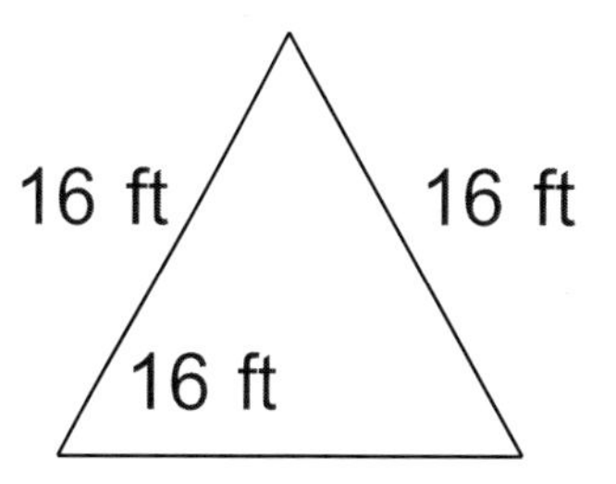

Page 76

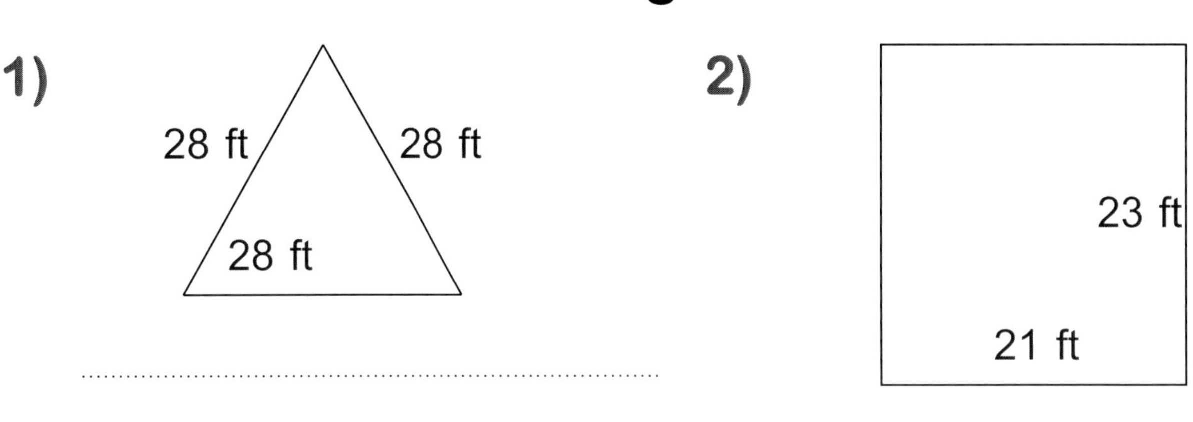
1)
28 ft
28 ft
28 ft
2)
23 ft
21 ft

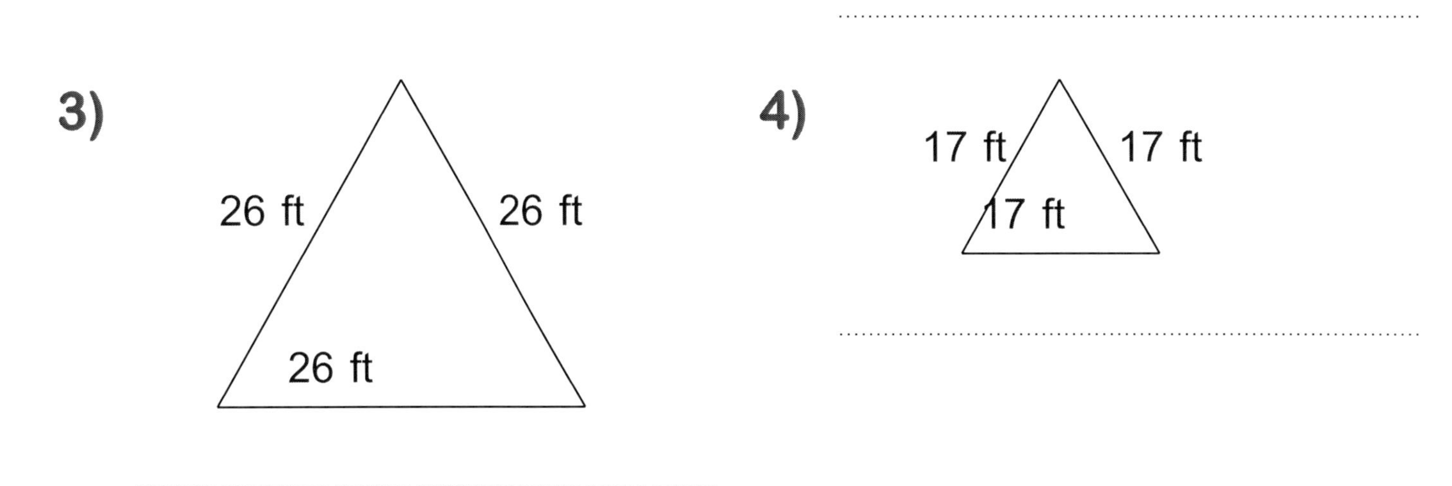
3)
26 ft
26 ft
26 ft
4)
17 ft
17 ft
17 ft

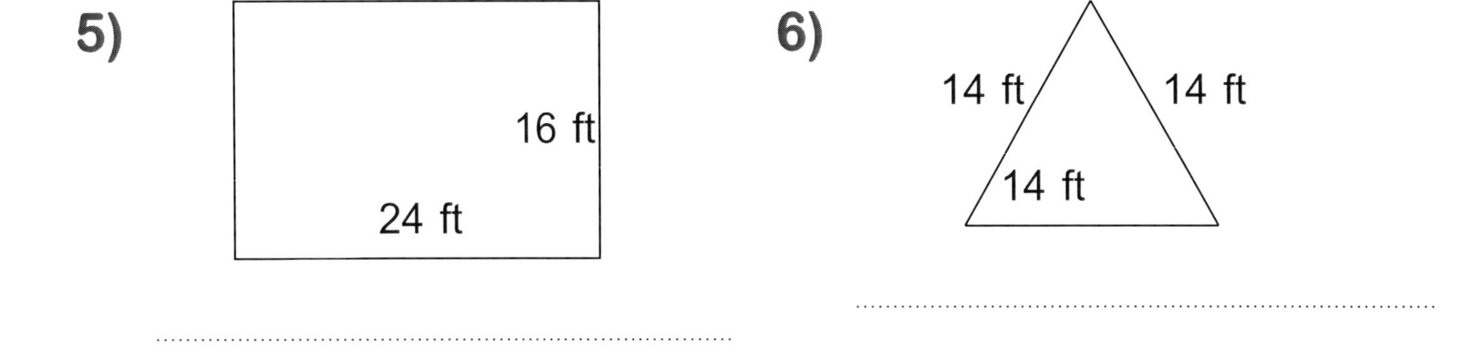
5)
16 ft
24 ft
6)
14 ft
14 ft
14 ft

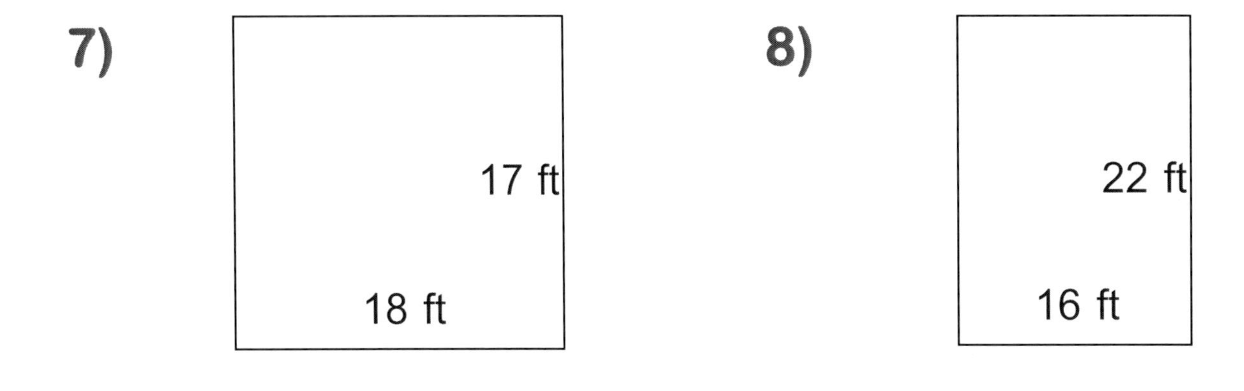
7)
17 ft
18 ft
8)
22 ft
16 ft

Page 77

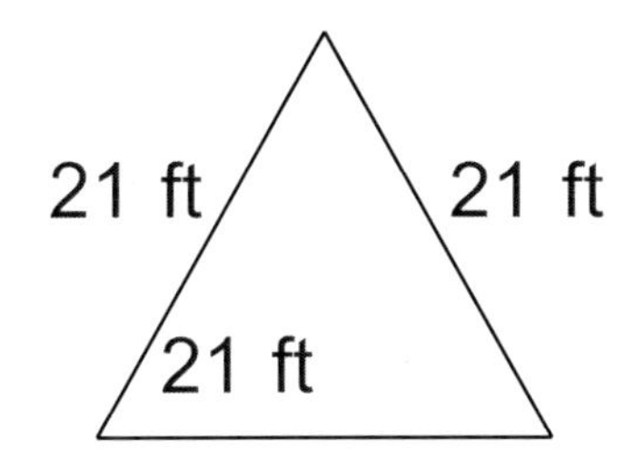

....

2)

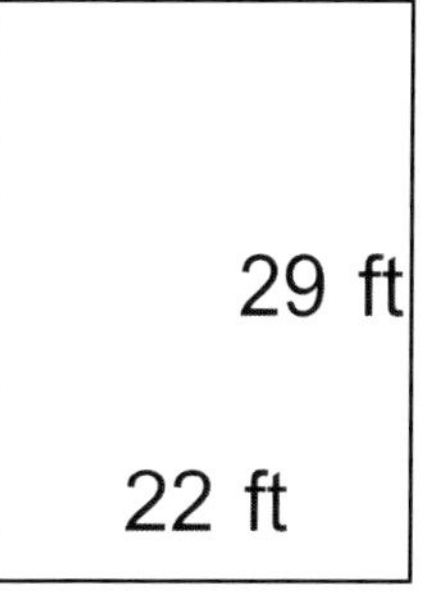

....

3)

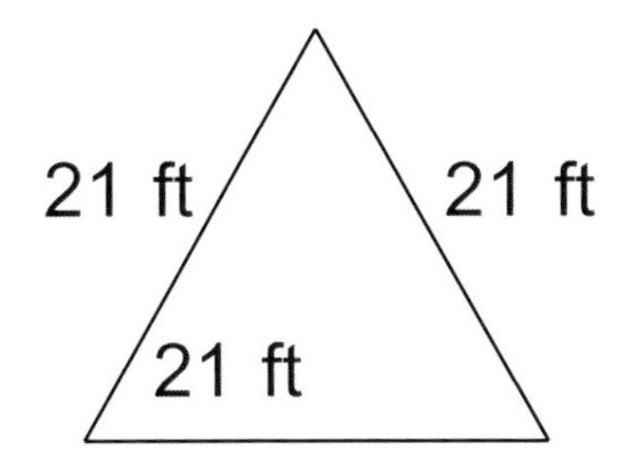

....

4)

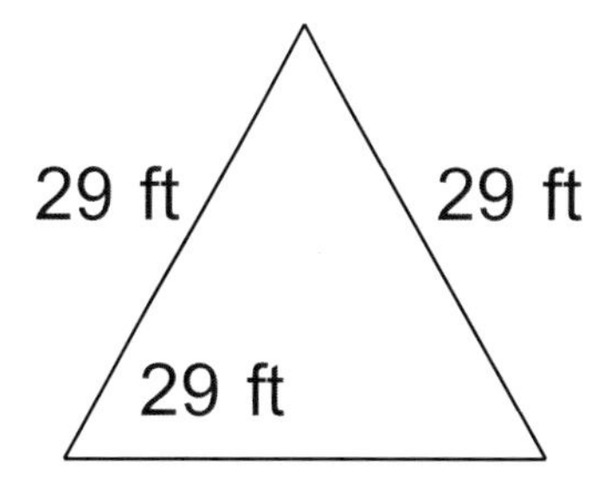

....

5)

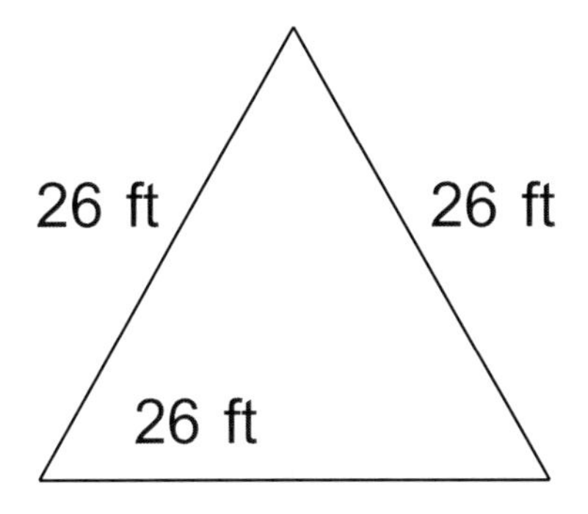

....

6)

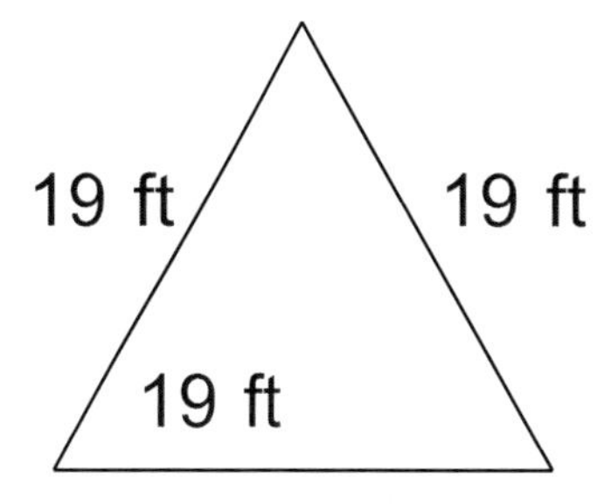

....

7)

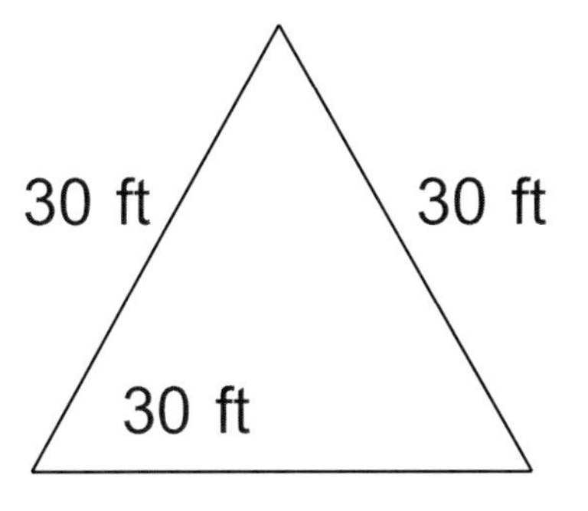

....

8)

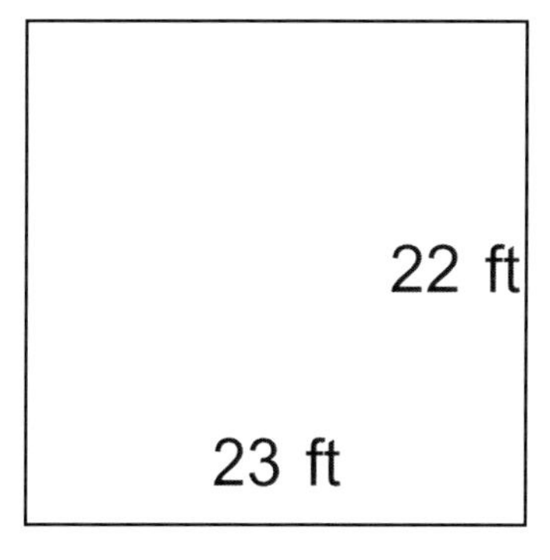

....

Page 78

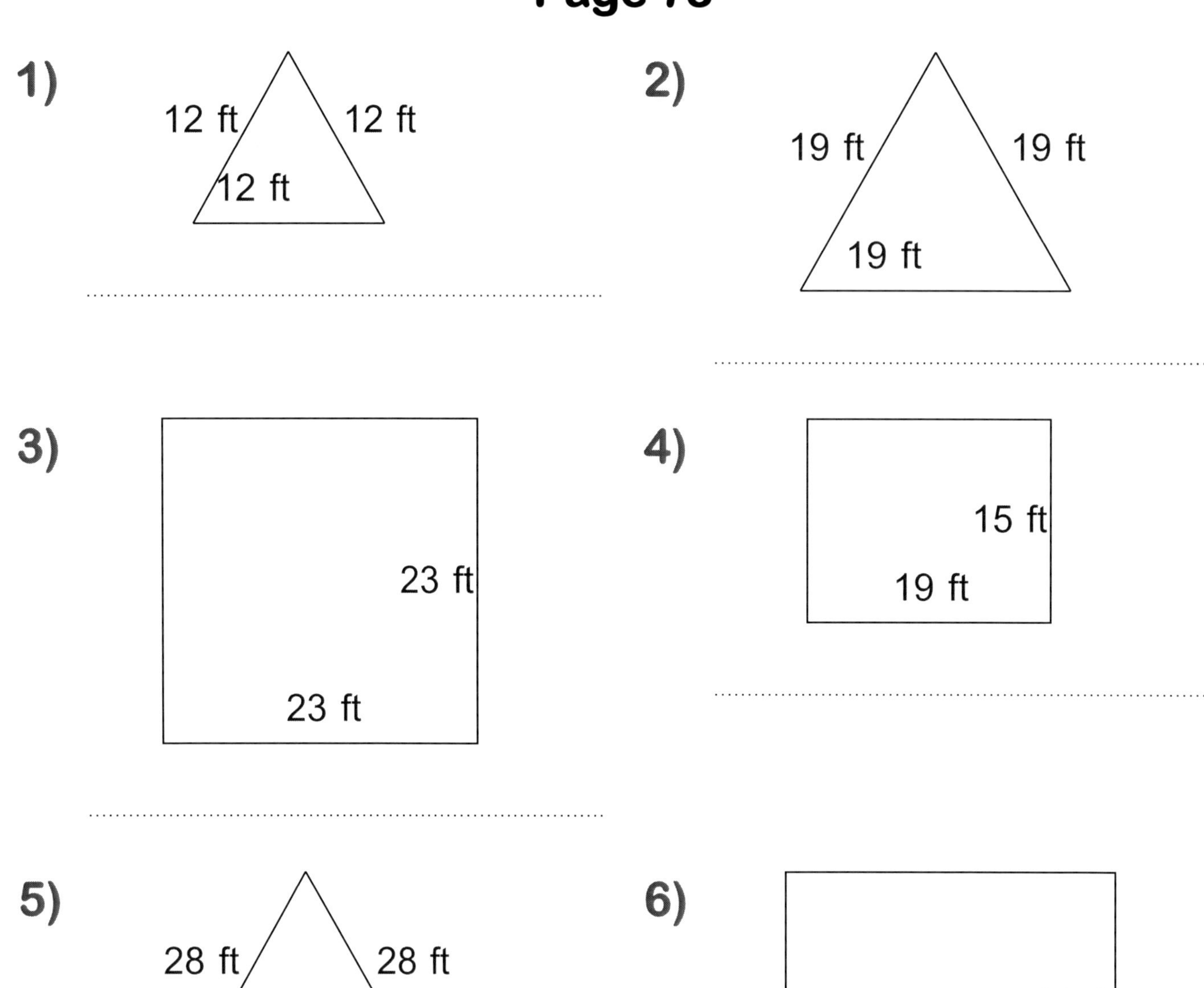
1)
12 ft
12 ft
12 ft
2)
19 ft
19 ft
19 ft
3)
23 ft
23 ft
4)
15 ft
19 ft

5)
28 ft
28 ft
28 ft
6)
28 ft
35 ft

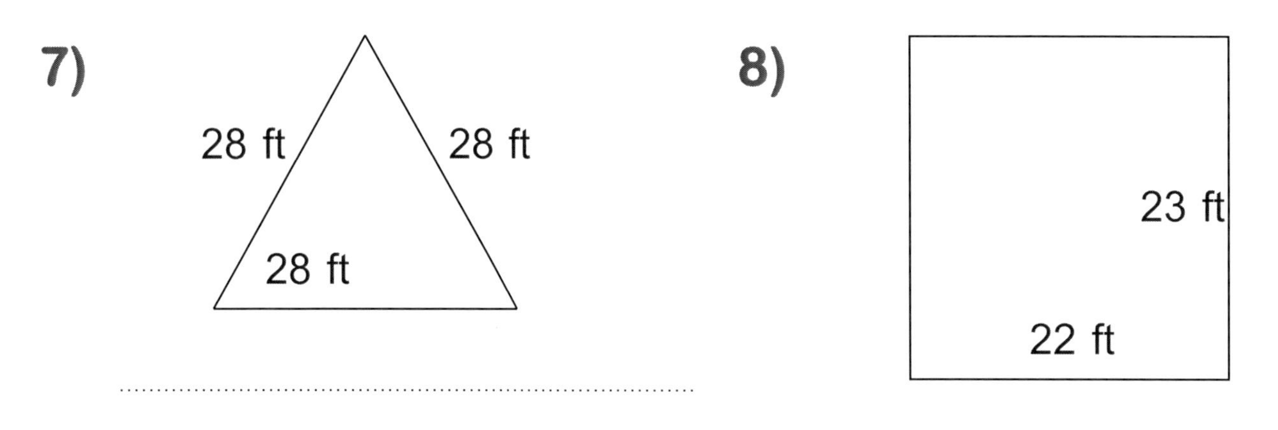
7)
28 ft
28 ft
28 ft
8)
23 ft
22 ft

Page 79

1)

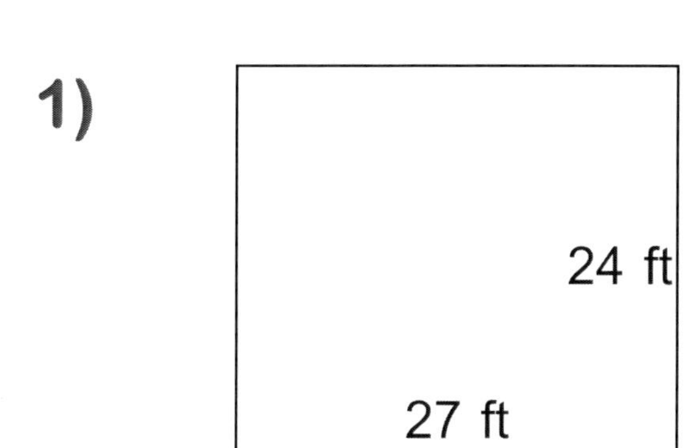

2)

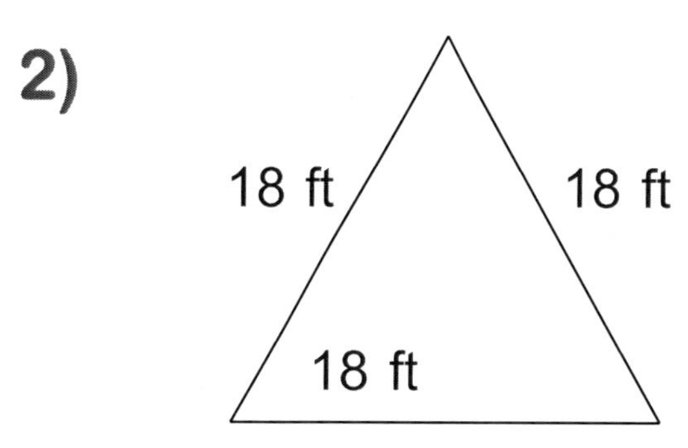

3)

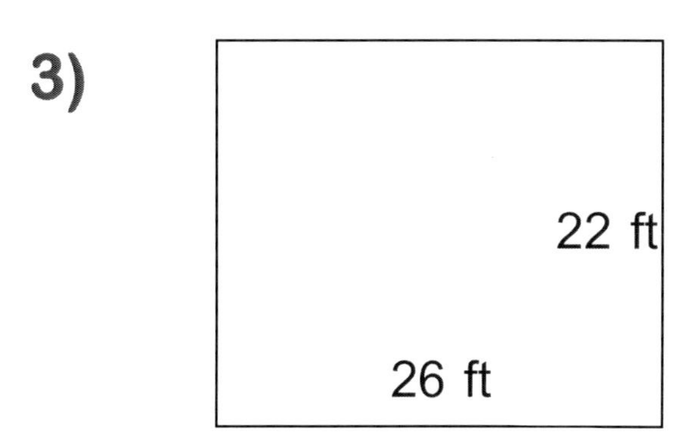

4)

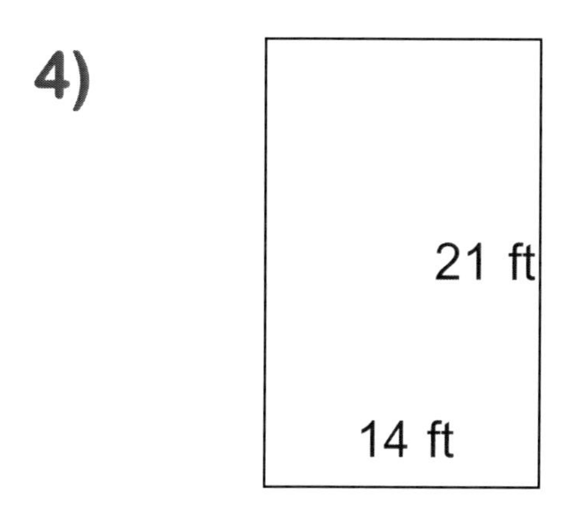

5)

12 ft
12 ft

6)

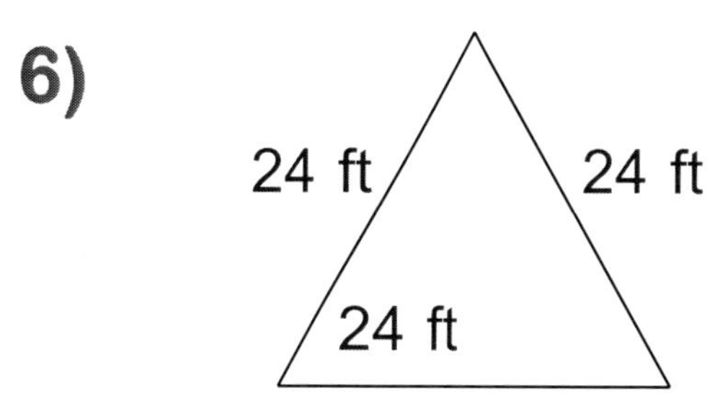

7)

28 ft
21 ft

8)

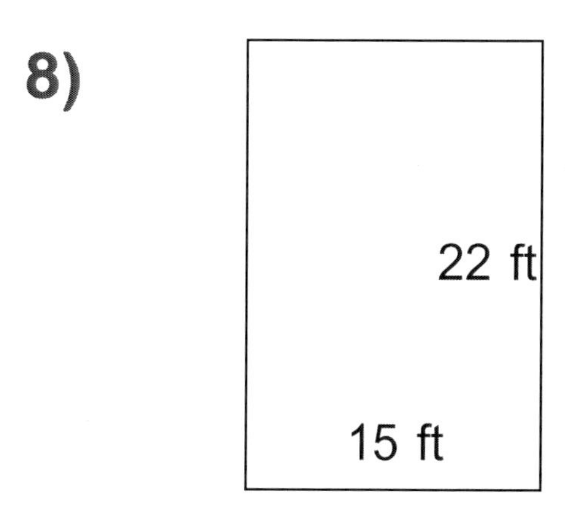

Page 80

1)

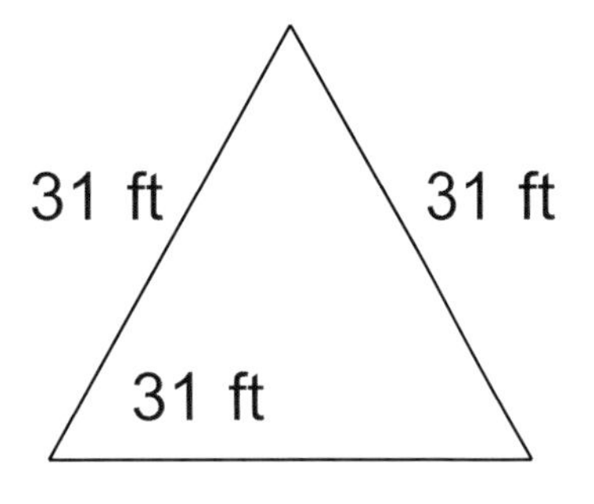

2)

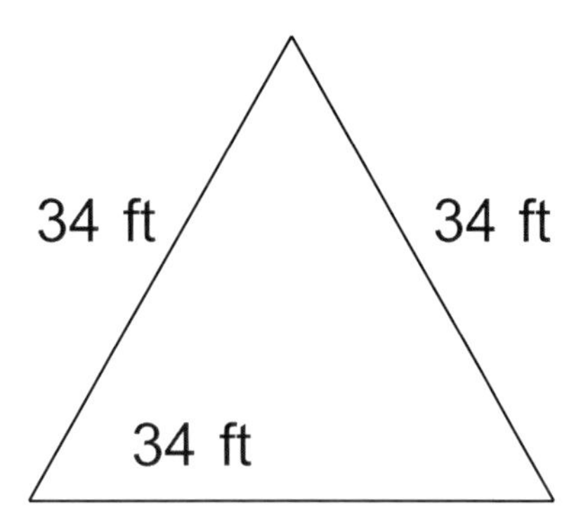

3)

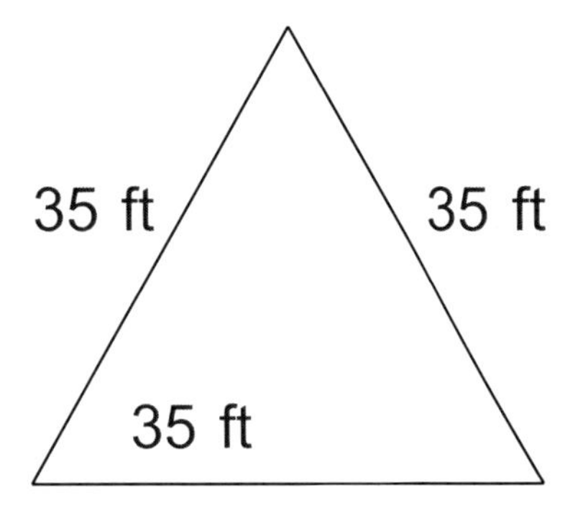

4)

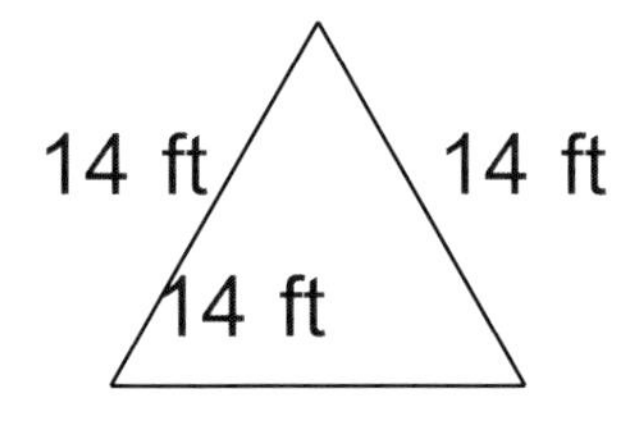

5)

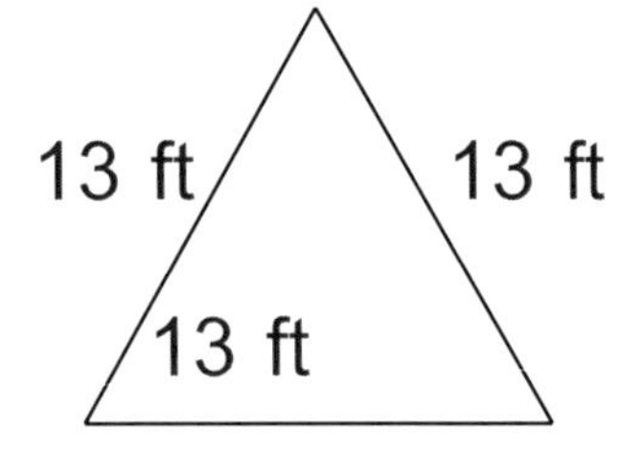

6)

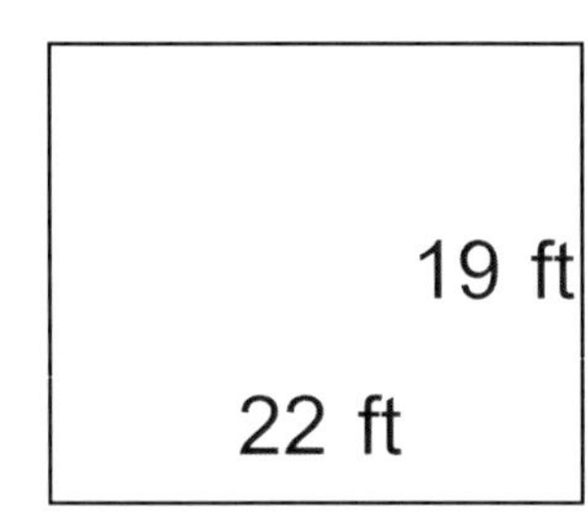

7)

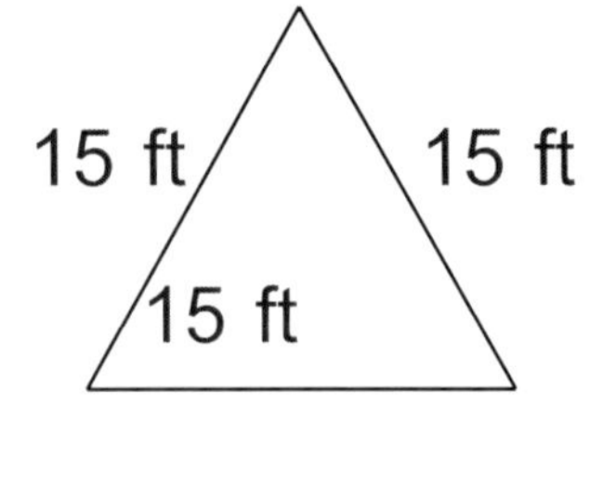

8)

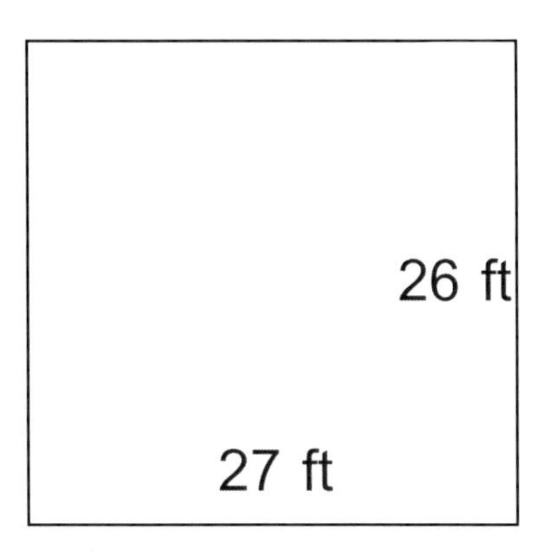

ANSWERS

Page 1

1) 19 + 10 29	2) 10 + 10 20	3) 14 + 13 27	4) 12 + 6 18
5) 11 + 16 27	6) 12 + 1 13	7) 14 + 15 29	8) 1 + 4 5
9) 2 + 2 4	10) 14 + 11 25	11) 6 + 11 17	12) 11 + 5 16
13) 10 + 1 11	14) 16 + 12 28	15) 4 + 5 9	16) 4 + 2 6
17) 3 + 15 18	18) 15 + 13 28	19) 10 + 18 28	20) 6 + 10 16

Page 2

1) 2 + 14 16	2) 5 + 1 6	3) 6 + 11 17	4) 7 + 10 17
5) 13 + 14 27	6) 2 + 17 19	7) 4 + 4 8	8) 3 + 4 7
9) 11 + 5 16	10) 11 + 7 18	11) 4 + 10 14	12) 14 + 14 28
13) 2 + 10 12	14) 10 + 19 29	15) 17 + 10 27	16) 7 + 12 19
17) 3 + 10 13	18) 10 + 6 16	19) 13 + 4 17	20) 4 + 20 24

Page 3

1) 11 + 13 24	2) 4 + 12 16	3) 5 + 2 7	4) 17 + 10 27
5) 10 + 15 25	6) 13 + 12 25	7) 10 + 18 28	8) 7 + 20 27
9) 4 + 3 7	10) 11 + 12 23	11) 17 + 12 29	12) 13 + 5 18
13) 5 + 4 9	14) 7 + 10 17	15) 11 + 2 13	16) 13 + 11 24
17) 16 + 1 17	18) 15 + 3 18	19) 4 + 14 18	20) 15 + 13 28

Page 4

1) 1 + 3 4	2) 14 + 12 26	3) 1 + 10 11	4) 16 + 11 27
5) 7 + 2 9	6) 15 + 12 27	7) 11 + 10 21	8) 2 + 6 8
9) 1 + 2 3	10) 19 + 10 29	11) 8 + 10 18	12) 3 + 3 6
13) 12 + 13 25	14) 6 + 10 16	15) 20 + 5 25	16) 13 + 10 23
17) 7 + 11 18	18) 13 + 11 24	19) 3 + 5 8	20) 3 + 6 9

Page 5

1) 7 + 1 = 8	2) 2 + 14 = 16	3) 11 + 10 = 21	4) 6 + 12 = 18
5) 15 + 11 = 26	6) 2 + 15 = 17	7) 10 + 5 = 15	8) 10 + 3 = 13
9) 11 + 1 = 12	10) 2 + 13 = 15	11) 16 + 2 = 18	12) 13 + 16 = 29
13) 1 + 4 = 5	14) 3 + 16 = 19	15) 2 + 12 = 14	16) 7 + 11 = 18
17) 10 + 7 = 17	18) 11 + 5 = 16	19) 10 + 12 = 22	20) 19 + 20 = 39

Page 6

1) 3 + 3 = 6	2) 4 + 1 = 5	3) 20 + 9 = 29	4) 17 + 2 = 19
5) 2 + 3 = 5	6) 15 + 11 = 26	7) 2 + 15 = 17	8) 10 + 5 = 15
9) 13 + 13 = 26	10) 1 + 1 = 2	11) 7 + 12 = 19	12) 15 + 4 = 19
13) 11 + 16 = 27	14) 19 + 10 = 29	15) 10 + 17 = 27	16) 12 + 12 = 24
17) 13 + 20 = 33	18) 4 + 15 = 19	19) 16 + 2 = 18	20) 15 + 13 = 28

Page 7

1) 3 + 15 = 18	2) 3 + 6 = 9	3) 14 + 1 = 15	4) 13 + 15 = 28
5) 16 + 12 = 28	6) 3 + 14 = 17	7) 5 + 12 = 17	8) 11 + 2 = 13
9) 1 + 1 = 2	10) 4 + 14 = 18	11) 1 + 11 = 12	12) 3 + 10 = 13
13) 2 + 17 = 19	14) 2 + 7 = 9	15) 4 + 10 = 14	16) 11 + 6 = 17
17) 14 + 13 = 27	18) 4 + 5 = 9	19) 14 + 2 = 16	20) 15 + 1 = 16

Page 8

1) 11 + 7 = 18	2) 11 + 2 = 13	3) 5 + 12 = 17	4) 19 + 10 = 29
5) 17 + 11 = 28	6) 13 + 13 = 26	7) 4 + 13 = 17	8) 10 + 18 = 28
9) 16 + 10 = 26	10) 7 + 11 = 18	11) 8 + 11 = 19	12) 3 + 15 = 18
13) 10 + 13 = 23	14) 17 + 2 = 19	15) 11 + 14 = 25	16) 8 + 10 = 18
17) 15 + 3 = 18	18) 15 + 4 = 19	19) 4 + 2 = 6	20) 18 + 10 = 28

Page 9

1) 1 + 13 = 14	2) 1 + 16 = 17	3) 14 + 14 = 28	4) 10 + 5 = 15
5) 2 + 4 = 6	6) 15 + 13 = 28	7) 18 + 10 = 28	8) 11 + 5 = 16
9) 11 + 6 = 17	10) 4 + 14 = 18	11) 10 + 12 = 22	12) 12 + 13 = 25
13) 11 + 7 = 18	14) 6 + 11 = 17	15) 8 + 10 = 18	16) 2 + 1 = 3
17) 14 + 11 = 25	18) 11 + 17 = 28	19) 11 + 2 = 13	20) 13 + 5 = 18

Page 10

1) 14 + 4 = 18	2) 18 + 10 = 28	3) 6 + 10 = 16	4) 16 + 10 = 26
5) 3 + 5 = 8	6) 14 + 11 = 25	7) 1 + 8 = 9	8) 7 + 1 = 8
9) 11 + 15 = 26	10) 4 + 4 = 8	11) 3 + 15 = 18	12) 16 + 20 = 36
13) 10 + 12 = 22	14) 15 + 3 = 18	15) 2 + 6 = 8	16) 15 + 2 = 17
17) 14 + 20 = 34	18) 17 + 11 = 28	19) 12 + 13 = 25	20) 5 + 3 = 8

Page 11

1) 10 + 6 = 16	2) 15 + 3 = 18	3) 14 + 4 = 18	4) 16 + 11 = 27
5) 1 + 12 = 13	6) 3 + 6 = 9	7) 15 + 4 = 19	8) 3 + 20 = 23
9) 1 + 15 = 16	10) 14 + 13 = 27	11) 11 + 4 = 15	12) 12 + 12 = 24
13) 11 + 7 = 18	14) 15 + 13 = 28	15) 11 + 11 = 22	16) 3 + 3 = 6
17) 17 + 12 = 29	18) 11 + 3 = 14	19) 5 + 20 = 25	20) 4 + 20 = 24

Page 12

1) 9 marbles were in the basket. 7 are red and the rest are green. How many marbles are green?

2

2) Ellen has 8 more plums than Amy. Amy has 6 plums. How many plums does Ellen have?

14

3) Jake has 1 ball and Adam has 7 balls. How many balls do Jake and Adam have together?

8

4) Some pears were in the basket. 3 more pears were added to the basket. Now there are 5 pears. How many pears were in the basket before more pears were added?

2

5) 6 apples were in the basket. More apples were added to the basket. Now there are 11 apples. How many apples were added to the basket?

5

Page 13

1) Amy has 9 more apples than Sharon. Sharon has 5 apples. How many apples does Amy have?

   14

2) Paul has 10 balls and Steven has 7 balls. How many balls do Paul and Steven have together?

   17

3) 8 pears were in the basket. 5 are red and the rest are green. How many pears are green?

   3

4) Some peaches were in the basket. 10 more peaches were added to the basket. Now there are 11 peaches. How many peaches were in the basket before more peaches were added?

   1

5) 4 plums were in the basket. More plums were added to the basket. Now there are 9 plums. How many plums were added to the basket?

   5

Page 14

1) 10 balls were in the basket. 9 are red and the rest are green. How many balls are green?

   1

2) 4 red apples and 6 green apples are in the basket. How many apples are in the basket?

   10

3) Some marbles were in the basket. 5 more marbles were added to the basket. Now there are 14 marbles. How many marbles were in the basket before more marbles were added?

   9

4) 2 plums were in the basket. More plums were added to the basket. Now there are 4 plums. How many plums were added to the basket?

   2

5) 5 oranges are in the basket. 9 more oranges are put in the basket. How many oranges are in the basket now?

   14

Page 15

1) Allan has 3 peaches and Brian has 8 peaches. How many peaches do Allan and Brian have together?

   11

2) Some balls were in the basket. 7 more balls were added to the basket. Now there are 12 balls. How many balls were in the basket before more balls were added?

   5

3) 8 marbles are in the basket. 5 more marbles are put in the basket. How many marbles are in the basket now?

   13

4) 3 apples were in the basket. More apples were added to the basket. Now there are 9 apples. How many apples were added to the basket?

   6

5) 12 plums were in the basket. 10 are red and the rest are green. How many plums are green?

   2

Page 16

1) $9 - 3 = 6$
2) $23 - 11 = 12$
3) $16 - 5 = 11$
4) $9 - 6 = 3$
5) $12 - 4 = 8$
6) $12 - 9 = 3$
7) $7 - 4 = 3$
8) $27 - 12 = 15$
9) $6 - 4 = 2$
10) $27 - 4 = 23$
11) $2 - 2 = 0$
12) $12 - 3 = 9$
13) $20 - 5 = 15$
14) $1 - 1 = 0$
15) $8 - 2 = 6$
16) $11 - 7 = 4$
17) $3 - 1 = 2$
18) $13 - 4 = 9$
19) $10 - 9 = 1$
20) $21 - 17 = 4$

Page 17

1) 3 - 2 = 1	2) 19 - 4 = 15	3) 11 - 2 = 9	4) 26 - 12 = 14
5) 1 - 1 = 0	6) 25 - 12 = 13	7) 15 - 11 = 4	8) 20 - 2 = 18
9) 20 - 7 = 13	10) 12 - 8 = 4	11) 11 - 4 = 7	12) 16 - 3 = 13
13) 6 - 4 = 2	14) 25 - 4 = 21	15) 4 - 3 = 1	16) 6 - 3 = 3
17) 27 - 19 = 8	18) 7 - 5 = 2	19) 4 - 2 = 2	20) 23 - 13 = 10

Page 18

1) 22 - 10 = 12	2) 17 - 12 = 5	3) 19 - 6 = 13	4) 7 - 3 = 4
5) 24 - 16 = 8	6) 6 - 6 = 0	7) 25 - 20 = 5	8) 13 - 5 = 8
9) 8 - 6 = 2	10) 18 - 6 = 12	11) 15 - 15 = 0	12) 17 - 4 = 13
13) 6 - 5 = 1	14) 3 - 2 = 1	15) 6 - 2 = 4	16) 26 - 15 = 11
17) 19 - 1 = 18	18) 18 - 10 = 8	19) 1 - 1 = 0	20) 28 - 12 = 16

Page 19

1) 23 - 7 = 16	2) 11 - 3 = 8	3) 20 - 18 = 2	4) 12 - 7 = 5
5) 8 - 6 = 2	6) 2 - 1 = 1	7) 2 - 2 = 0	8) 4 - 2 = 2
9) 3 - 2 = 1	10) 11 - 4 = 7	11) 19 - 2 = 17	12) 5 - 1 = 4
13) 4 - 4 = 0	14) 11 - 8 = 3	15) 23 - 19 = 4	16) 17 - 3 = 14
17) 19 - 8 = 11	18) 16 - 5 = 11	19) 11 - 7 = 4	20) 7 - 4 = 3

Page 20

1) 3 - 2 = 1	2) 21 - 4 = 17	3) 30 - 1 = 29	4) 13 - 8 = 5
5) 16 - 7 = 9	6) 9 - 4 = 5	7) 22 - 7 = 15	8) 2 - 2 = 0
9) 16 - 13 = 3	10) 12 - 5 = 7	11) 11 - 2 = 9	12) 26 - 18 = 8
13) 28 - 14 = 14	14) 29 - 16 = 13	15) 5 - 4 = 1	16) 6 - 2 = 4
17) 29 - 9 = 20	18) 13 - 3 = 10	19) 17 - 2 = 15	20) 15 - 12 = 3

Page 21

1) 6 - 3 3	2) 21 - 4 17	3) 13 - 5 8	4) 12 - 11 1
5) 23 - 11 12	6) 6 - 5 1	7) 2 - 2 0	8) 12 - 8 4
9) 10 - 2 8	10) 3 - 2 1	11) 19 - 14 5	12) 10 - 6 4
13) 8 - 5 3	14) 11 - 7 4	15) 5 - 2 3	16) 9 - 3 6
17) 26 - 12 14	18) 7 - 1 6	19) 12 - 9 3	20) 15 - 10 5

Page 22

1) 11 - 1 10	2) 1 - 1 0	3) 18 - 6 12	4) 5 - 2 3
5) 27 - 10 17	6) 7 - 4 3	7) 2 - 1 1	8) 27 - 18 9
9) 30 - 15 15	10) 24 - 10 14	11) 8 - 7 1	12) 29 - 10 19
13) 16 - 6 10	14) 8 - 5 3	15) 19 - 9 10	16) 12 - 8 4
17) 21 - 14 7	18) 28 - 19 9	19) 13 - 2 11	20) 19 - 7 12

Page 23

1) 13 - 11 2	2) 9 - 2 7	3) 13 - 10 3	4) 2 - 1 1
5) 24 - 12 12	6) 19 - 11 8	7) 7 - 4 3	8) 11 - 8 3
9) 22 - 13 9	10) 18 - 6 12	11) 25 - 11 14	12) 12 - 8 4
13) 23 - 7 16	14) 9 - 9 0	15) 15 - 14 1	16) 5 - 3 2
17) 18 - 15 3	18) 3 - 3 0	19) 11 - 2 9	20) 24 - 2 22

Page 24

1) 26 - 16 10	2) 12 - 3 9	3) 11 - 3 8	4) 18 - 13 5
5) 12 - 12 0	6) 23 - 3 20	7) 1 - 1 0	8) 22 - 16 6
9) 30 - 14 16	10) 27 - 17 10	11) 4 - 3 1	12) 14 - 10 4
13) 16 - 15 1	14) 4 - 2 2	15) 19 - 18 1	16) 10 - 6 4
17) 16 - 13 3	18) 21 - 12 9	19) 7 - 5 2	20) 29 - 2 27

Page 25

1) $12 - 9 = 3$
2) $17 - 4 = 13$
3) $22 - 18 = 4$
4) $13 - 11 = 2$
5) $4 - 2 = 2$
6) $24 - 13 = 11$
7) $6 - 6 = 0$
8) $28 - 4 = 24$
9) $12 - 8 = 4$
10) $19 - 9 = 10$
11) $4 - 1 = 3$
12) $20 - 5 = 15$
13) $19 - 5 = 14$
14) $3 - 1 = 2$
15) $18 - 8 = 10$
16) $22 - 17 = 5$
17) $13 - 5 = 8$
18) $6 - 2 = 4$
19) $30 - 13 = 17$
20) $16 - 5 = 11$

Page 26

1) $12 - 5 = 7$
2) $5 - 4 = 1$
3) $11 - 7 = 4$
4) $26 - 11 = 15$
5) $7 - 4 = 3$
6) $29 - 11 = 18$
7) $28 - 7 = 21$
8) $20 - 3 = 17$
9) $16 - 8 = 8$
10) $10 - 3 = 7$
11) $24 - 19 = 5$
12) $20 - 8 = 12$
13) $27 - 7 = 20$
14) $16 - 9 = 7$
15) $29 - 3 = 26$
16) $21 - 1 = 20$
17) $28 - 17 = 11$
18) $22 - 15 = 7$
19) $10 - 10 = 0$
20) $27 - 2 = 25$

Page 27

1) 6 oranges are in the basket. 3 are red and the rest are green. How many oranges are green?

   3

2) 2 plums are in the basket. 2 plums are taken out of the basket. How many plums are in the basket now?

   0

3) Some balls were in the basket. 2 balls were taken from the basket. Now there is 1 ball. How many balls were in the basket before some of the balls were taken?

   3

4) 4 marbles were in the basket. Some of the marbles were removed from the basket. Now there is 1 marble. How many marbles were removed from the basket?

   3

5) Janet has 5 fewer pears than Michele. Michele has 10 pears. How many pears does Janet have?

   5

Page 28

1) Janet has 0 fewer peaches than Audrey. Audrey has 3 peaches. How many peaches does Janet have?

   3

2) Some balls were in the basket. 3 balls were taken from the basket. Now there is 1 ball. How many balls were in the basket before some of the balls were taken?

   4

3) 7 apples were in the basket. Some of the apples were removed from the basket. Now there are 4 apples. How many apples were removed from the basket?

   3

4) 4 pears are in the basket. 4 are red and the rest are green. How many pears are green?

   0

5) 4 marbles are in the basket. 4 marbles are taken out of the basket. How many marbles are in the basket now?

   0

Page 29

1) Brian has 6 balls. Allan has 9 balls. How many more balls does Allan have than Brian?

3

2) Michele has 0 fewer peaches than Marin. Marin has 6 peaches. How many peaches does Michele have?

6

3) 7 apples are in the basket. 3 are red and the rest are green. How many apples are green?

4

4) Some marbles were in the basket. 6 marbles were taken from the basket. Now there are 0 marbles. How many marbles were in the basket before some of the marbles were taken?

6

5) 5 pears were in the basket. Some of the pears were removed from the basket. Now there are 2 pears. How many pears were removed from the basket?

3

Page 30

1) 10 marbles are in the basket. 5 marbles are taken out of the basket. How many marbles are in the basket now?

5

2) Jake has 5 peaches. David has 5 peaches. How many more peaches does David have than Jake?

0

3) 10 balls were in the basket. Some of the balls were removed from the basket. Now there are 4 balls. How many balls were removed from the basket?

6

4) 3 apples are in the basket. 3 are red and the rest are green. How many apples are green?

0

5) Jennifer has 0 fewer pears than Ellen. Ellen has 5 pears. How many pears does Jennifer have?

5

Page 31

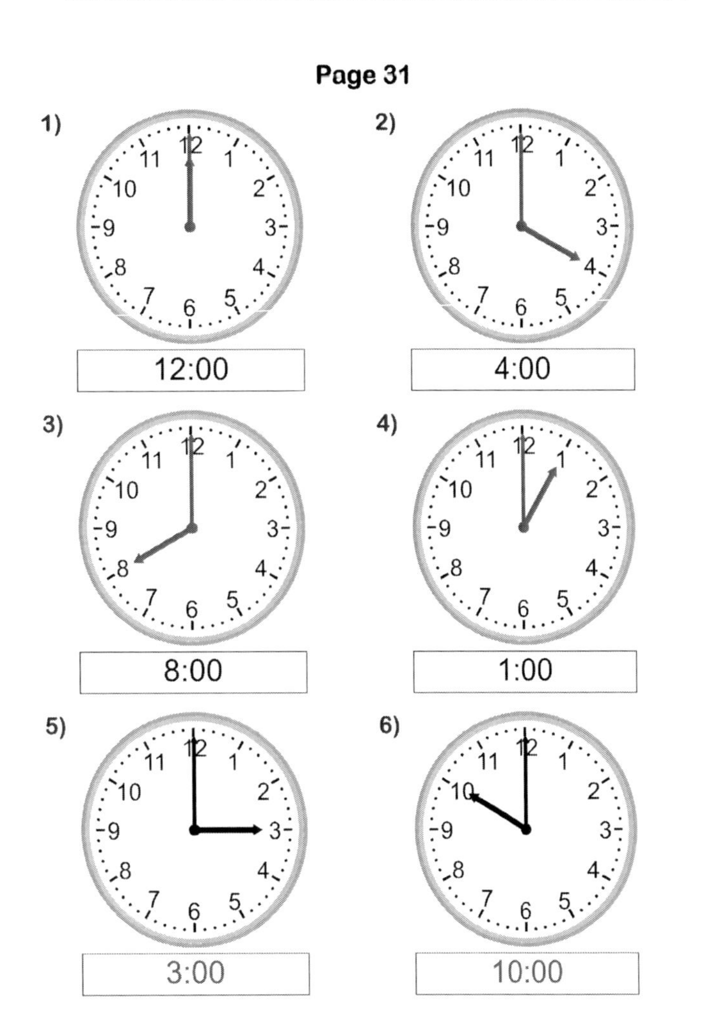

Page 32

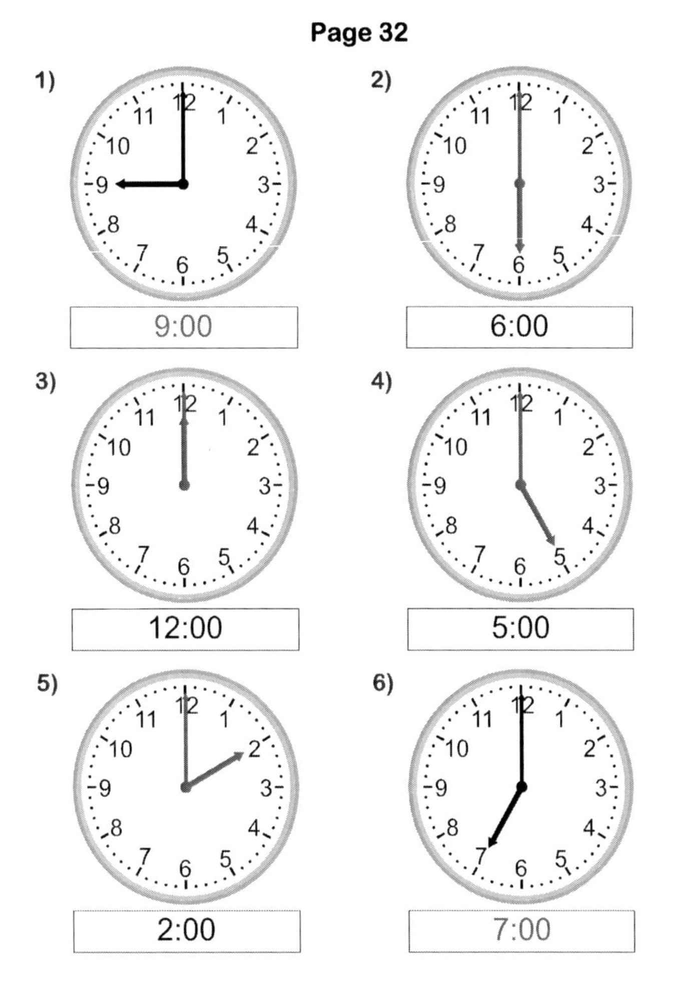

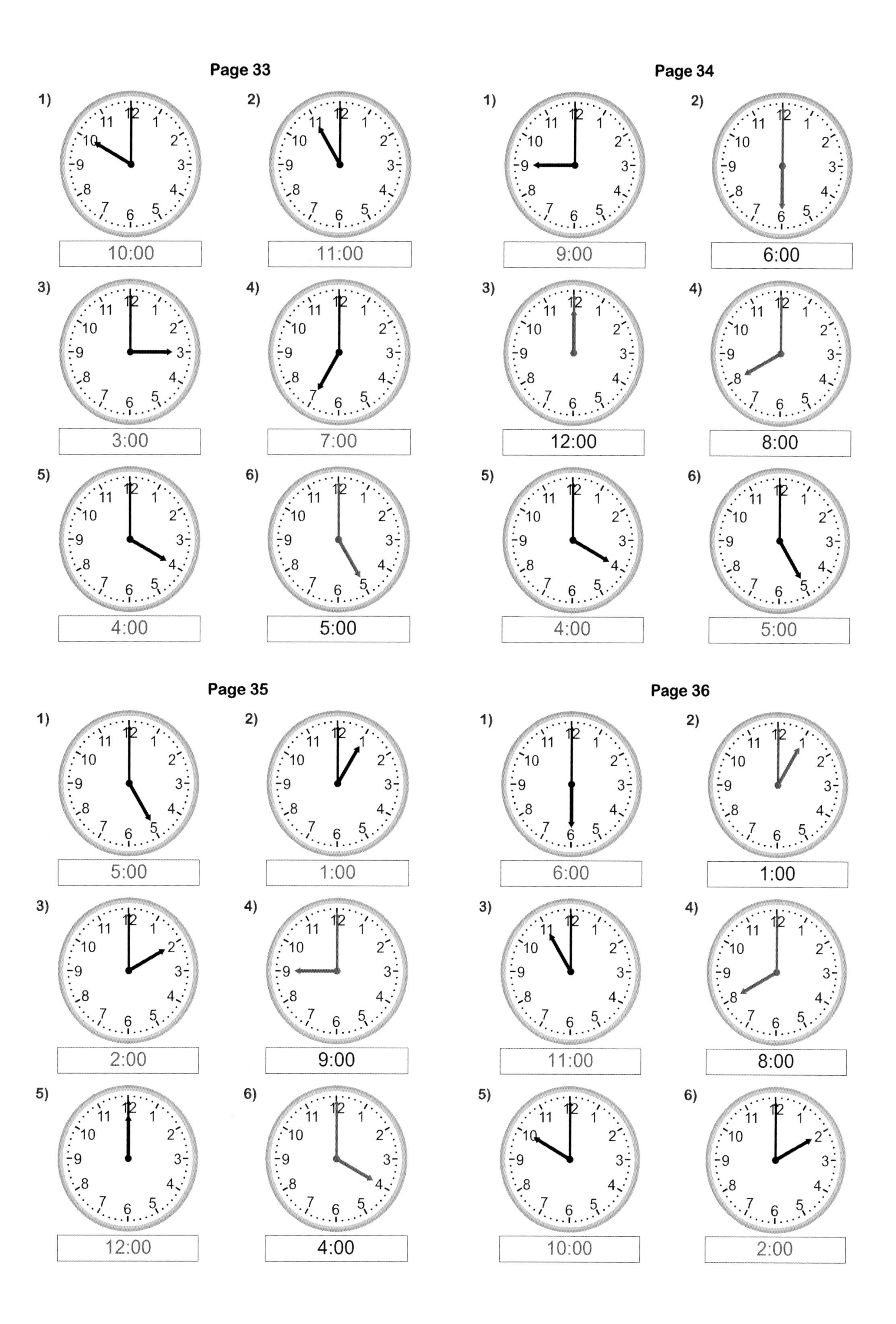

Page 33
1) 10:00
2) 11:00
3) 3:00
4) 7:00
5) 4:00
6) 5:00
Page 34
1) 9:00
2) 6:00
3) 12:00
4) 8:00
5) 4:00
6) 5:00
Page 35
1) 5:00
2) 1:00
3) 2:00
4) 9:00
5) 12:00
6) 4:00
Page 36
1) 6:00
2) 1:00
3) 11:00
4) 8:00
5) 10:00
6) 2:00

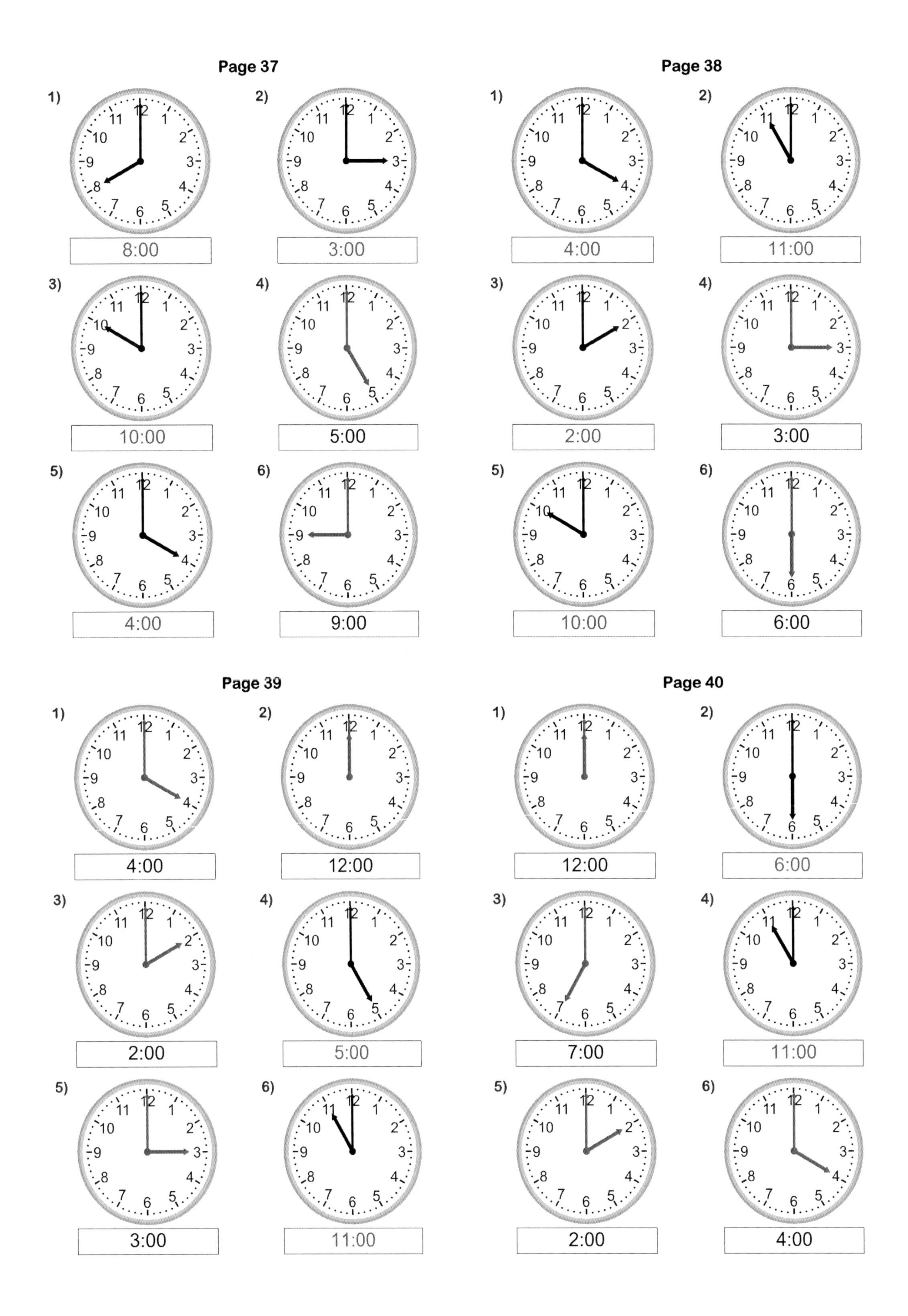
Page 37
1)
8:00
2)
3:00
3)
10:00
4)
5:00
5)
4:00
6)
9:00
Page 38
1)
4:00
2)
11:00
3)
2:00
4)
3:00
5)
10:00
6)
6:00
Page 39
1)
4:00
2)
12:00
3)
2:00
4)
5:00
5)
3:00
6)
11:00
Page 40
1)
12:00
2)
6:00
3)
7:00
4)
11:00
5)
2:00
6)
4:00

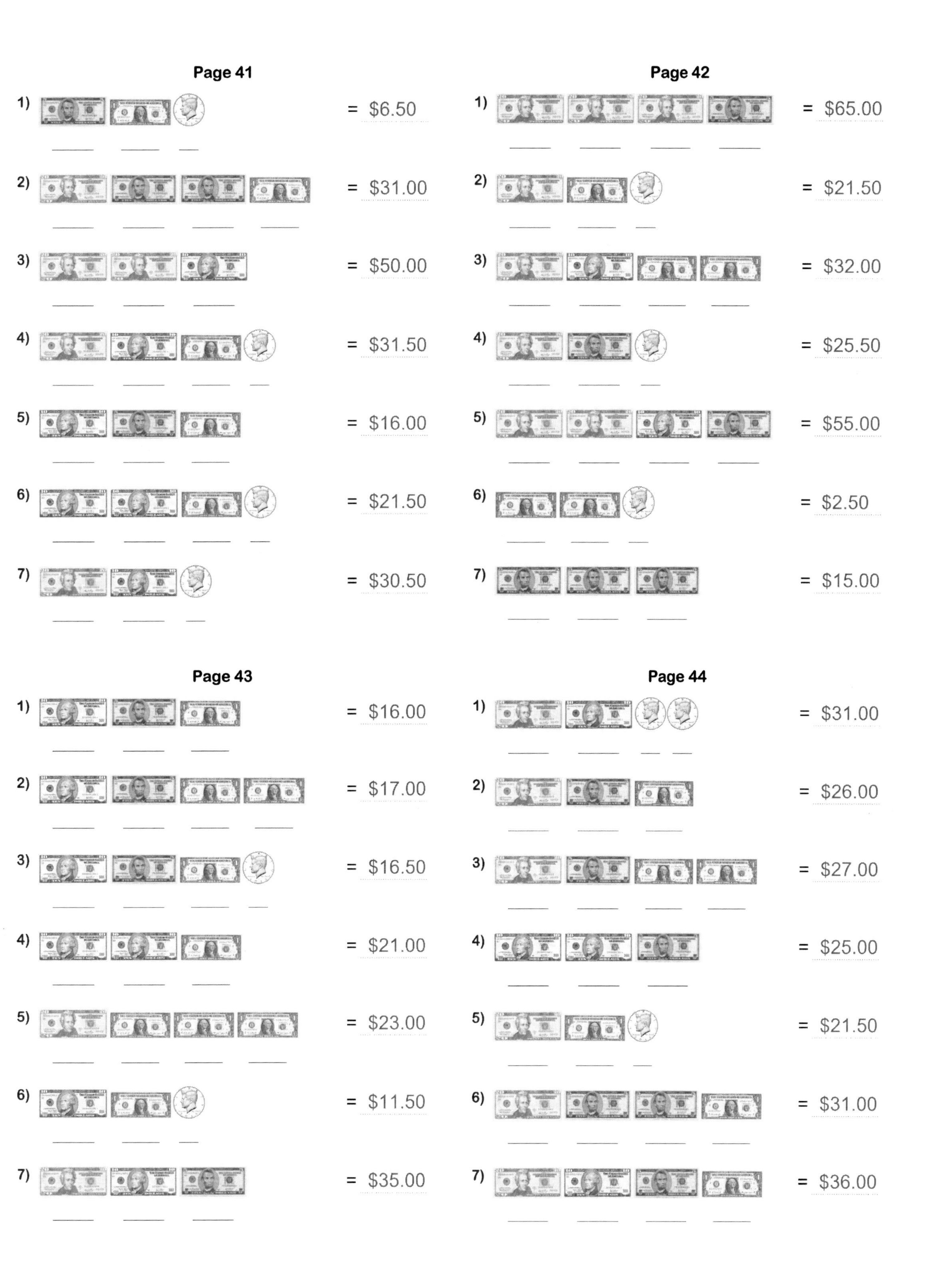

Page 41

1) = $6.50
2) = $31.00
3) = $50.00
4) = $31.50
5) = $16.00
6) = $21.50
7) = $30.50

Page 42

1) = $65.00
2) = $21.50
3) = $32.00
4) = $25.50
5) = $55.00
6) = $2.50
7) = $15.00

Page 43

1) = $16.00
2) = $17.00
3) = $16.50
4) = $21.00
5) = $23.00
6) = $11.50
7) = $35.00

Page 44

1) = $31.00
2) = $26.00
3) = $27.00
4) = $25.00
5) = $21.50
6) = $31.00
7) = $36.00

Page 45

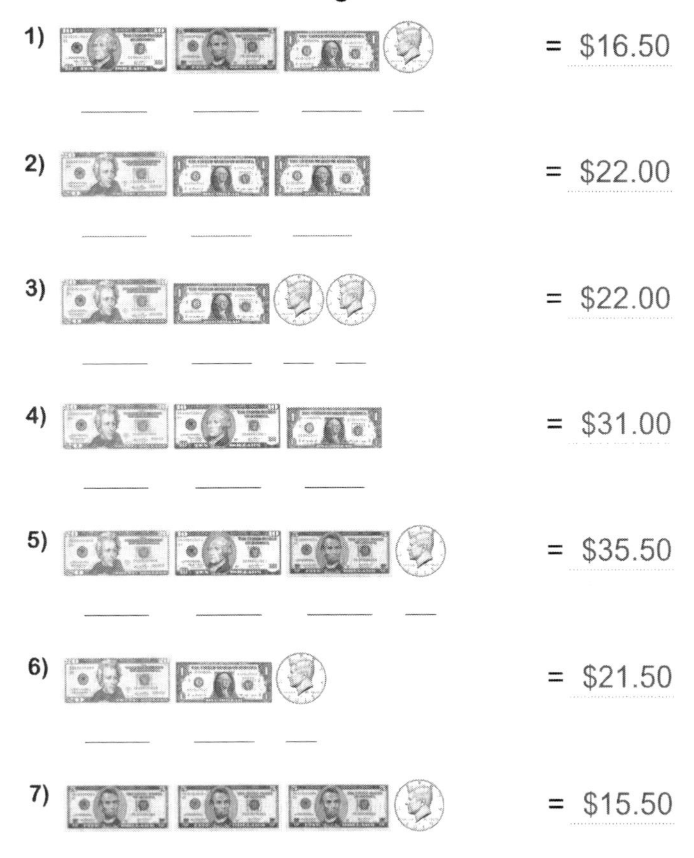

Page 46

hot dog = $1.00
order of French-fries = $1.00
hamburger = $2.00
deluxe cheeseburger = $3.00
cola = $1.00
ice cream cone = $1.00
milk shake = $2.00
taco = $2.00

1) $9.00 If Janet wanted to buy an ice cream cone and four tacos, how much money would she need?

2) $14.00 Amy wants to buy a deluxe cheeseburger, three colas, and four hamburgers. How much will it cost her?

3) $3.00 If Billy buys four ice cream cones and three orders of French-fries, how much change will he get back from $10.00?

4) $5.00 Sharon purchases five deluxe cheeseburgers, five hot dogs, and five orders of French-fries. If she had $30.00, how much money will she have left?

Page 47

hot dog = $1.00
order of French-fries = $1.00
hamburger = $2.00
deluxe cheeseburger = $3.00
cola = $1.00
ice cream cone = $1.00
milk shake = $2.00
taco = $2.00

1) $5.00 Jake purchases four orders of French-fries and a cola. How much change will he get back from $10.00?

2) $18.00 Paul wants to buy five tacos and four hamburgers. How much money will he need?

3) $21.00 If Adam wanted to buy a milk shake, two colas, five ice cream cones, two hamburgers, and four tacos, how much money would he need?

4) $32.00 Jennifer wants to buy two hot dogs, four tacos, five ice cream cones, two colas, and five deluxe cheeseburgers. How much will it cost her?

Page 48

hot dog = $1.00
order of French-fries = $0.00
hamburger = $2.00
deluxe cheeseburger = $3.00
cola = $1.00
ice cream cone = $1.00
milk shake = $2.00
taco = $2.00

1) $15.00 If Ellen wanted to buy five deluxe cheeseburgers, how much money would she need?

2) $8.00 What is the total cost of four milk shakes?

3) $21.00 If Jake wanted to buy four milk shakes, four colas, two orders of French-fries, three hot dogs, and three hamburgers, how much would he have to pay?

4) $18.00 Brian wants to buy four ice cream cones, two orders of French-fries, four tacos, and two deluxe cheeseburgers. How much will he have to pay?

Page 49

hot dog = $1.00
order of French-fries = $1.00
hamburger = $2.00
deluxe cheeseburger = $3.00
cola = $1.00
ice cream cone = $1.00
milk shake = $2.00
taco = $2.00

1) $25.00 If Steven wanted to buy four hamburgers, a deluxe cheeseburger, three tacos, and four milk shakes, how much would he have to pay?

2) $6.00 Ellen purchases four hamburgers and a cola. How much change will she get back from $15.00?

3) $22.00 If Jackie wanted to buy four orders of French-fries, two deluxe cheeseburgers, four hamburgers, and four hot dogs, how much would she have to pay?

4) $18.00 Sharon wants to buy three tacos, five milk shakes, and two colas. How much money will she need?

Page 50

hot dog = $1.00
order of French-fries = $1.00
hamburger = $2.00
deluxe cheeseburger = $3.00
cola = $1.00
ice cream cone = $1.00
milk shake = $2.00
taco = $2.00

1) $16.00 Adam wants to buy four deluxe cheeseburgers and two hamburgers. How much money will he need?

2) $4.00 Jackie wants to buy two tacos. How much money will she need?

3) $9.00 What is the total cost of a hamburger, three hot dogs, and four colas?

4) $3.00 Janet purchases four hamburgers and four orders of French-fries. If she had $15.00, how much money will she have left?

Page 51

1) $\frac{5}{8} > \frac{3}{8}$ 2) $\frac{3}{4} > \frac{1}{4}$ 3) $\frac{1}{5} < \frac{4}{5}$

4) $\frac{8}{6} < \frac{17}{6}$ 5) $\frac{15}{21} > \frac{3}{21}$ 6) $\frac{5}{9} < \frac{11}{9}$

7) $\frac{23}{10} > \frac{7}{10}$ 8) $\frac{1}{2} = \frac{1}{2}$ 9) $\frac{2}{6} < \frac{15}{6}$

10) $\frac{12}{8} > \frac{1}{8}$ 11) $\frac{4}{3} < \frac{7}{3}$ 12) $\frac{3}{4} > \frac{2}{4}$

13) $\frac{16}{7} = \frac{16}{7}$ 14) $\frac{8}{24} < \frac{54}{24}$ 15) $\frac{5}{8} < \frac{14}{8}$

16) $\frac{3}{2} < \frac{5}{2}$ 17) $\frac{40}{45} < \frac{120}{45}$ 18) $\frac{26}{10} > \frac{5}{10}$

19) $\frac{6}{5} > \frac{2}{5}$ 20) $\frac{4}{7} < \frac{5}{7}$ 21) $\frac{4}{10} < \frac{27}{10}$

Page 52

1) $\frac{5}{2} > \frac{1}{2}$ 2) $\frac{15}{9} > \frac{8}{9}$ 3) $\frac{2}{14} < \frac{12}{14}$

4) $\frac{15}{30} > \frac{13}{30}$ 5) $\frac{3}{4} > \frac{1}{4}$ 6) $\frac{2}{5} > \frac{1}{5}$

7) $\frac{14}{20} < \frac{49}{20}$ 8) $\frac{6}{7} < \frac{13}{7}$ 9) $\frac{24}{30} < \frac{76}{30}$

10) $\frac{8}{3} > \frac{2}{3}$ 11) $\frac{6}{9} > \frac{3}{9}$ 12) $\frac{3}{15} < \frac{44}{15}$

13) $\frac{3}{6} < \frac{4}{6}$ 14) $\frac{7}{4} > \frac{1}{4}$ 15) $\frac{11}{6} > \frac{1}{6}$

16) $\frac{2}{8} < \frac{7}{8}$ 17) $\frac{4}{20} < \frac{11}{20}$ 18) $\frac{1}{3} < \frac{7}{3}$

19) $\frac{15}{7} > \frac{13}{7}$ 20) $\frac{1}{2} = \frac{1}{2}$ 21) $\frac{3}{4} = \frac{3}{4}$

Page 53

1) $\frac{1}{2} < \frac{3}{2}$ 2) $\frac{10}{8} < \frac{14}{8}$ 3) $\frac{1}{10} < \frac{3}{10}$

4) $\frac{6}{4} > \frac{2}{4}$ 5) $\frac{3}{8} < \frac{22}{8}$ 6) $\frac{10}{20} < \frac{34}{20}$

7) $\frac{20}{9} > \frac{11}{9}$ 8) $\frac{6}{21} > \frac{5}{21}$ 9) $\frac{8}{10} > \frac{7}{10}$

10) $\frac{21}{10} > \frac{1}{10}$ 11) $\frac{23}{8} > \frac{7}{8}$ 12) $\frac{2}{6} < \frac{10}{6}$

13) $\frac{1}{2} = \frac{1}{2}$ 14) $\frac{7}{3} > \frac{5}{3}$ 15) $\frac{10}{4} > \frac{6}{4}$

16) $\frac{21}{10} > \frac{9}{10}$ 17) $\frac{15}{35} < \frac{27}{35}$ 18) $\frac{6}{9} = \frac{6}{9}$

19) $\frac{6}{36} < \frac{80}{36}$ 20) $\frac{2}{5} < \frac{7}{5}$ 21) $\frac{3}{6} = \frac{3}{6}$

Page 54

1) $\frac{5}{8} < \frac{11}{8}$ 2) $\frac{27}{10} > \frac{1}{10}$ 3) $\frac{6}{5} > \frac{3}{5}$

4) $\frac{14}{9} = \frac{14}{9}$ 5) $\frac{8}{28} < \frac{20}{28}$ 6) $\frac{10}{6} > \frac{2}{6}$

7) $\frac{7}{10} > \frac{1}{10}$ 8) $\frac{6}{8} < \frac{19}{8}$ 9) $\frac{12}{48} > \frac{11}{48}$

10) $\frac{5}{2} > \frac{3}{2}$ 11) $\frac{10}{15} < \frac{43}{15}$ 12) $\frac{3}{2} > \frac{1}{2}$

13) $\frac{12}{5} > \frac{4}{5}$ 14) $\frac{3}{7} < \frac{19}{7}$ 15) $\frac{24}{36} < \frac{40}{36}$

16) $\frac{9}{30} < \frac{74}{30}$ 17) $\frac{3}{4} > \frac{2}{4}$ 18) $\frac{5}{3} > \frac{2}{3}$

19) $\frac{10}{40} > \frac{4}{40}$ 20) $\frac{17}{6} = \frac{17}{6}$ 21) $\frac{7}{8} > \frac{6}{8}$

Page 55

1) $\frac{1}{5} < \frac{11}{5}$ 2) $\frac{7}{3} > \frac{1}{3}$ 3) $\frac{6}{12} = \frac{6}{12}$

4) $\frac{4}{8} = \frac{4}{8}$ 5) $\frac{7}{4} > \frac{3}{4}$ 6) $\frac{6}{14} < \frac{10}{14}$

7) $\frac{4}{18} < \frac{16}{18}$ 8) $\frac{16}{10} > \frac{3}{10}$ 9) $\frac{36}{48} > \frac{32}{48}$

10) $\frac{24}{9} > \frac{13}{9}$ 11) $\frac{12}{30} < \frac{49}{30}$ 12) $\frac{4}{3} > \frac{1}{3}$

13) $\frac{8}{10} < \frac{29}{10}$ 14) $\frac{8}{6} < \frac{14}{6}$ 15) $\frac{3}{4} > \frac{2}{4}$

16) $\frac{5}{7} = \frac{5}{7}$ 17) $\frac{3}{2} = \frac{3}{2}$ 18) $\frac{7}{3} > \frac{2}{3}$

19) $\frac{6}{7} < \frac{13}{7}$ 20) $\frac{4}{5} < \frac{9}{5}$ 21) $\frac{4}{8} > \frac{1}{8}$

Page 56

1) $\frac{1}{3} =$ 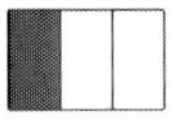 2) $\frac{4}{5} =$

3) $\frac{1}{2} =$  4) $\frac{2}{5} =$

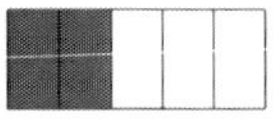

5) $\frac{3}{4} =$ 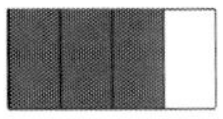 6) $\frac{1}{4} =$

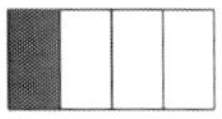

7) $\frac{2}{3} =$ 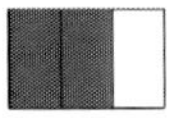 8) $\frac{3}{5} =$

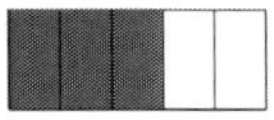

9) $\frac{1}{5} =$ 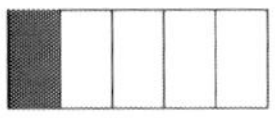 10) $\frac{2}{4} =$

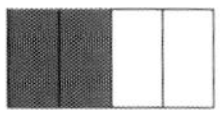

11) $\frac{3}{5} =$ 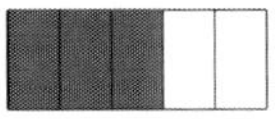 12) $\frac{3}{4} =$

13) $\frac{3}{5} =$ 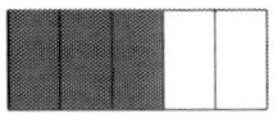 14) $\frac{2}{5} =$

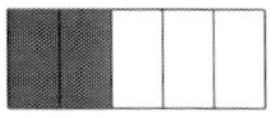

Page 57

1) $\frac{1}{3}$ = 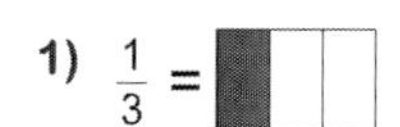

2) $\frac{1}{2}$ =

3) $\frac{2}{5}$ =

4) $\frac{3}{4}$ =

5) $\frac{2}{3}$ =

6) $\frac{2}{4}$ =

7) $\frac{1}{5}$ =

8) $\frac{3}{5}$ =

9) $\frac{1}{4}$ =

10) $\frac{4}{5}$ =

11) $\frac{1}{3}$ =

12) $\frac{1}{5}$ =

13) $\frac{1}{4}$ =

14) $\frac{1}{4}$ =

Page 58

1) $\frac{2}{3}$ =

2) $\frac{1}{2}$ =

3) $\frac{2}{4}$ =

4) $\frac{1}{5}$ =

5) $\frac{1}{4}$ =

6) $\frac{1}{3}$ =

7) $\frac{3}{4}$ =

8) $\frac{2}{5}$ =

9) $\frac{3}{5}$ =

10) $\frac{4}{5}$ =

11) $\frac{2}{3}$ =

12) $\frac{1}{2}$ =

13) $\frac{1}{2}$ =

14) $\frac{3}{5}$ =

Page 59

1) $\frac{3}{4}$ =

2) $\frac{1}{5}$ =

3) $\frac{1}{3}$ =

4) $\frac{1}{2}$ =

5) $\frac{3}{5}$ =

6) $\frac{2}{3}$ =

7) $\frac{2}{5}$ =

8) $\frac{2}{4}$ =

9) $\frac{4}{5}$ =

10) $\frac{1}{4}$ =

11) $\frac{2}{3}$ =

12) $\frac{4}{5}$ =

13) $\frac{3}{5}$ =

14) $\frac{1}{3}$ =

Page 60

1) $\frac{1}{5}$ =

2) $\frac{2}{5}$ =

3) $\frac{1}{2}$ =

4) $\frac{1}{3}$ =

5) $\frac{3}{4}$ =

6) $\frac{2}{4}$ =

7) $\frac{2}{3}$ =

8) $\frac{4}{5}$ =

9) $\frac{3}{5}$ =

10) $\frac{1}{4}$ =

11) $\frac{4}{5}$ =

12) $\frac{2}{5}$ =

13) $\frac{1}{2}$ =

14) $\frac{1}{2}$ =

Page 61

1) 8 ft = 2.438 m
2) 13 ft = 3.962 m
3) 18 ft = 5.486 m
4) 12 ft = 3.658 m
5) 16 ft = 4.877 m
6) 12 ft = 3.658 m
7) 13 ft = 3.962 m
8) 9 ft = 2.743 m
9) 10 ft = 3.048 m
10) 9 ft = 2.743 m
11) 16 ft = 4.877 m
12) 8 ft = 2.438 m
13) 2 ft = 0.610 m
14) 1 ft = 0.305 m
15) 7 ft = 2.134 m
16) 4 ft = 1.219 m
17) 6 ft = 1.829 m
18) 16 ft = 4.877 m
19) 19 ft = 5.791 m
20) 5 ft = 1.524 m

Page 62

1) 16 ft = 4.877 m
2) 8 ft = 2.438 m
3) 19 ft = 5.791 m
4) 15 ft = 4.572 m
5) 6 ft = 1.829 m
6) 3 ft = 0.914 m
7) 13 ft = 3.962 m
8) 4 ft = 1.219 m
9) 5 ft = 1.524 m
10) 9 ft = 2.743 m
11) 18 ft = 5.486 m
12) 12 ft = 3.658 m
13) 14 ft = 4.267 m
14) 9 ft = 2.743 m
15) 17 ft = 5.182 m
16) 14 ft = 4.267 m
17) 6 ft = 1.829 m
18) 6 ft = 1.829 m
19) 8 ft = 2.438 m
20) 4 ft = 1.219 m

Page 63

1) 14 ft = 4.267 m
2) 3 ft = 0.914 m
3) 3 ft = 0.914 m
4) 6 ft = 1.829 m
5) 9 ft = 2.743 m
6) 3 ft = 0.914 m
7) 20 ft = 6.096 m
8) 17 ft = 5.182 m
9) 12 ft = 3.658 m
10) 1 ft = 0.305 m
11) 7 ft = 2.134 m
12) 7 ft = 2.134 m
13) 12 ft = 3.658 m
14) 6 ft = 1.829 m
15) 13 ft = 3.962 m
16) 9 ft = 2.743 m
17) 9 ft = 2.743 m
18) 7 ft = 2.134 m
19) 15 ft = 4.572 m
20) 5 ft = 1.524 m

Page 64

1) 11 ft = 3.353 m
2) 3 ft = 0.914 m
3) 12 ft = 3.658 m
4) 11 ft = 3.353 m
5) 8 ft = 2.438 m
6) 8 ft = 2.438 m
7) 12 ft = 3.658 m
8) 5 ft = 1.524 m
9) 5 ft = 1.524 m
10) 6 ft = 1.829 m
11) 11 ft = 3.353 m
12) 20 ft = 6.096 m
13) 19 ft = 5.791 m
14) 14 ft = 4.267 m
15) 11 ft = 3.353 m
16) 14 ft = 4.267 m
17) 7 ft = 2.134 m
18) 14 ft = 4.267 m
19) 7 ft = 2.134 m
20) 4 ft = 1.219 m

Page 65

1) 3 ft = 0.914 m
2) 16 ft = 4.877 m
3) 12 ft = 3.658 m
4) 4 ft = 1.219 m
5) 3 ft = 0.914 m
6) 12 ft = 3.658 m
7) 1 ft = 0.305 m
8) 9 ft = 2.743 m
9) 17 ft = 5.182 m
10) 10 ft = 3.048 m
11) 6 ft = 1.829 m
12) 19 ft = 5.791 m
13) 11 ft = 3.353 m
14) 4 ft = 1.219 m
15) 13 ft = 3.962 m
16) 12 ft = 3.658 m
17) 10 ft = 3.048 m
18) 12 ft = 3.658 m
19) 18 ft = 5.486 m
20) 2 ft = 0.610 m

Page 66

1) 6 in = 0 ft 6 in ft
2) 14 ft = 4 yd 2 ft yd
3) 8 in = 0 yd 8 in yd
4) 8 ft = 2 yd 2 ft yd
5) 17 in = 1 ft 5 in ft
6) 16 in = 1 ft 4 in ft
7) 20 ft = 6 yd 2 ft yd
8) 11 in = 0 yd 11 in yd
9) 5 ft = 1 yd 2 ft yd
10) 7 in = 0 ft 7 in ft
11) 15 ft = 5 yd yd
12) 1 in = 0 yd 1 in yd
13) 5 in = 0 yd 5 in yd
14) 18 in = 1 ft 6 in ft
15) 13 in = 1 ft 1 in ft
16) 11 ft = 3 yd 2 ft yd
17) 17 ft = 5 yd 2 ft yd
18) 4 in = 0 yd 4 in yd
19) 19 ft = 6 yd 1 ft yd
20) 13 in = 0 yd 13 in yd

Page 67

1) 5 ft = 1 yd 2 ft yd
2) 9 ft = 3 yd yd
3) 8 in = 0 ft 8 in ft
4) 8 ft = 2 yd 2 ft yd
5) 16 ft = 5 yd 1 ft yd
6) 3 ft = 1 yd yd
7) 15 ft = 5 yd yd
8) 14 ft = 4 yd 2 ft yd
9) 11 ft = 3 yd 2 ft yd
10) 2 ft = 0 yd 2 ft yd
11) 14 in = 0 yd 14 in yd
12) 19 in = 1 ft 7 in ft
13) 15 in = 1 ft 3 in ft
14) 4 in = 0 ft 4 in ft
15) 15 in = 0 yd 15 in yd
16) 12 in = 1 ft ft
17) 6 in = 0 yd 6 in yd
18) 13 in = 0 yd 13 in yd
19) 6 ft = 2 yd yd
20) 19 ft = 6 yd 1 ft yd

Page 68

1) 6 in = 0 ft 6 in ft
2) 17 ft = 5 yd 2 ft yd
3) 16 in = 1 ft 4 in ft
4) 4 ft = 1 yd 1 ft yd
5) 6 in = 0 yd 6 in yd
6) 18 ft = 6 yd yd
7) 2 in = 0 yd 2 in yd
8) 6 ft = 2 yd yd
9) 1 ft = 0 yd 1 ft yd
10) 19 ft = 6 yd 1 ft yd
11) 2 ft = 0 yd 2 ft yd
12) 14 ft = 4 yd 2 ft yd
13) 15 ft = 5 yd yd
14) 8 in = 0 ft 8 in ft
15) 7 in = 0 ft 7 in ft
16) 4 in = 0 ft 4 in ft
17) 9 ft = 3 yd yd
18) 12 in = 0 yd 12 in yd
19) 12 in = 1 ft ft
20) 17 in = 1 ft 5 in ft

Page 69

1) 8 ft = 2 yd 2 ft yd
2) 12 ft = 4 yd yd
3) 4 in = 0 yd 4 in yd
4) 14 ft = 4 yd 2 ft yd
5) 13 in = 0 yd 13 in yd
6) 14 in = 1 ft 2 in ft
7) 16 ft = 5 yd 1 ft yd
8) 9 ft = 3 yd yd
9) 5 in = 0 yd 5 in yd
10) 10 in = 0 ft 10 in ft
11) 2 ft = 0 yd 2 ft yd
12) 18 in = 0 yd 18 in yd
13) 15 in = 1 ft 3 in ft
14) 15 ft = 5 yd yd
15) 11 ft = 3 yd 2 ft yd
16) 6 in = 0 yd 6 in yd
17) 19 in = 1 ft 7 in ft
18) 20 in = 1 ft 8 in ft
19) 7 ft = 2 yd 1 ft yd
20) 10 in = 0 yd 10 in yd

Page 70

1) 17 ft = 5 yd 2 ft yd
2) 12 ft = 4 yd yd
3) 10 ft = 3 yd 1 ft yd
4) 10 in = 0 ft 10 in ft
5) 14 ft = 4 yd 2 ft yd
6) 5 in = 0 ft 5 in ft
7) 13 ft = 4 yd 1 ft yd
8) 9 in = 0 yd 9 in yd
9) 11 ft = 3 yd 2 ft yd
10) 4 ft = 1 yd 1 ft yd
11) 7 ft = 2 yd 1 ft yd
12) 7 in = 0 yd 7 in yd
13) 8 ft = 2 yd 2 ft yd
14) 6 ft = 2 yd yd
15) 10 in = 0 yd 10 in yd
16) 2 in = 0 ft 2 in ft
17) 2 in = 0 yd 2 in yd
18) 4 in = 0 ft 4 in ft
19) 17 in = 1 ft 5 in ft
20) 5 in = 0 yd 5 in yd

Page 71

1) 22 ft, 19 ft

P = 82 ft A = 418 ft²

2) 17 ft, 10 ft

P = 54 ft A = 170 ft²

3) 22 ft, 20 ft

P = 84 ft A = 440 ft²

4)

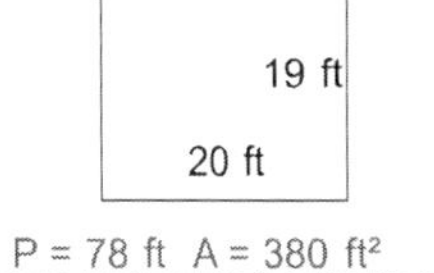

P = 78 ft A = 380 ft²

5) 28 ft, 25 ft

P = 106 ft A = 700 ft²

6)

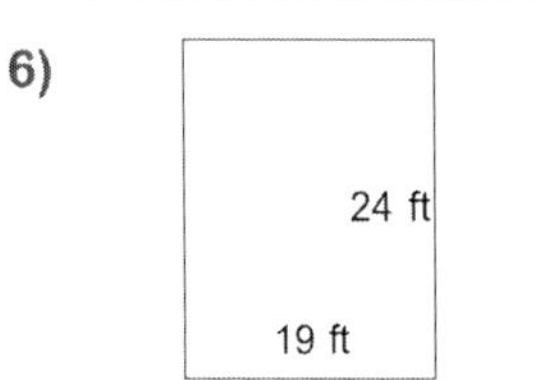

P = 86 ft A = 456 ft²

7)

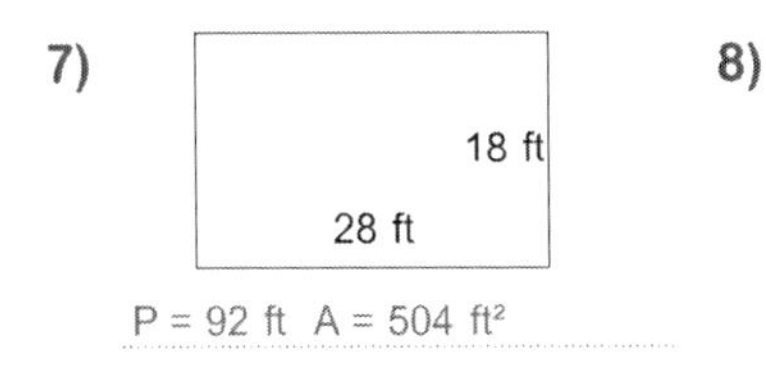

P = 92 ft A = 504 ft²

8) 30 ft, 18 ft

P = 96 ft A = 540 ft²

Page 72

1) 20 ft, 30 ft

P = 100 ft A = 600 ft²

2)

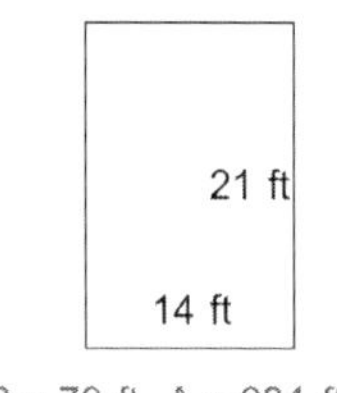

P = 70 ft A = 294 ft²

3) 21 ft, 28 ft

P = 98 ft A = 588 ft²

4)

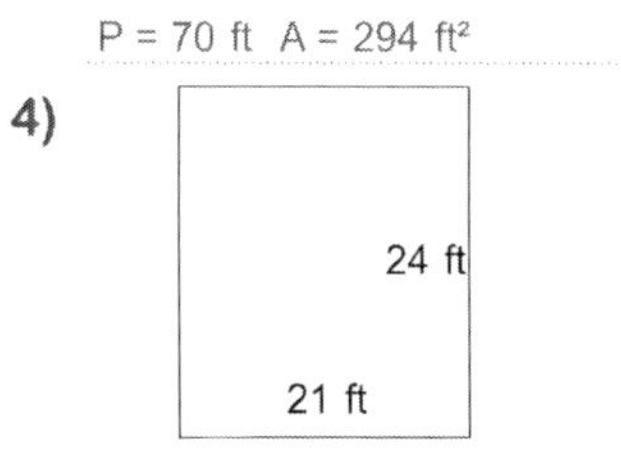

P = 90 ft A = 504 ft²

5) 35 ft, 36 ft

P = 142 ft A = 1,260 ft²

6)

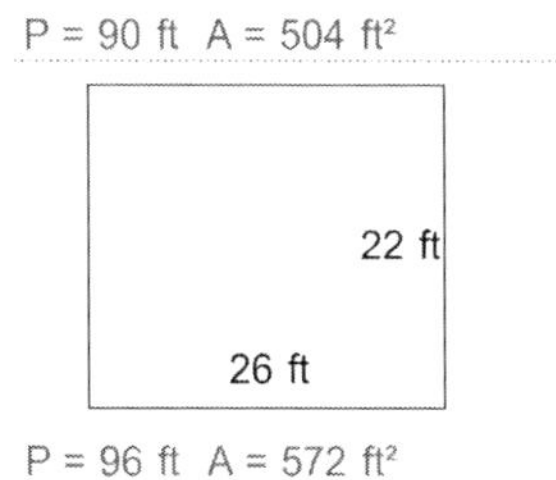

P = 96 ft A = 572 ft²

7) 26 ft, 24 ft

P = 100 ft A = 624 ft²

8) 34 ft, 34 ft

P = 136 ft A = 1,156 ft²

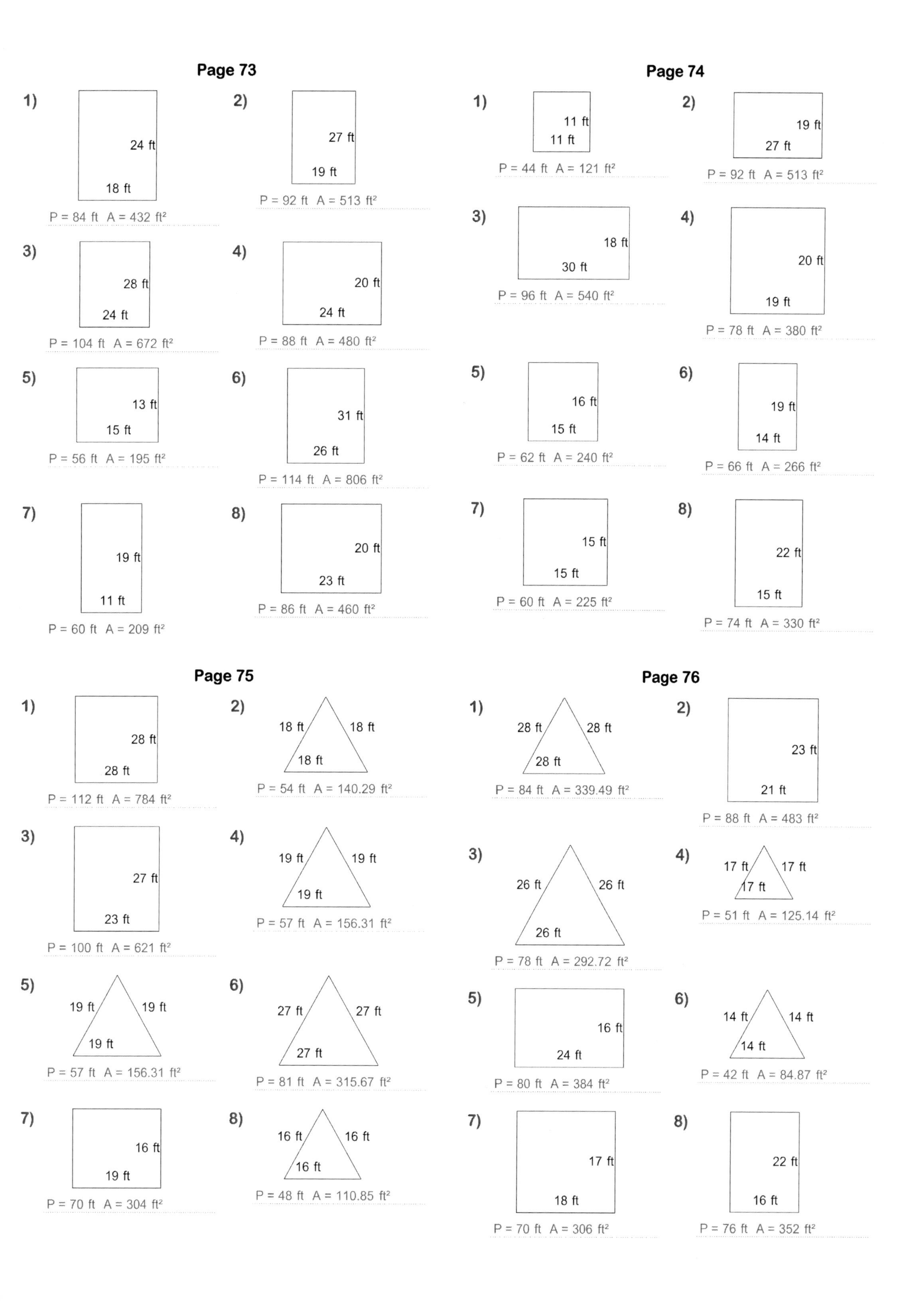
Page 73
1)
24 ft
18 ft
P = 84 ft A = 432 ft²
2)
27 ft
19 ft
P = 92 ft A = 513 ft²
3)
28 ft
24 ft
P = 104 ft A = 672 ft²
4)
20 ft
24 ft
P = 88 ft A = 480 ft²
5)
13 ft
15 ft
P = 56 ft A = 195 ft²
6)
31 ft
26 ft
P = 114 ft A = 806 ft²
7)
19 ft
11 ft
P = 60 ft A = 209 ft²
8)
20 ft
23 ft
P = 86 ft A = 460 ft²
Page 74
1)
11 ft
11 ft
P = 44 ft A = 121 ft²
2)
19 ft
27 ft
P = 92 ft A = 513 ft²
3)
18 ft
30 ft
P = 96 ft A = 540 ft²
4)
20 ft
19 ft
P = 78 ft A = 380 ft²
5)
16 ft
15 ft
P = 62 ft A = 240 ft²
6)
19 ft
14 ft
P = 66 ft A = 266 ft²
7)
15 ft
15 ft
P = 60 ft A = 225 ft²
8)
22 ft
15 ft
P = 74 ft A = 330 ft²
Page 75
1)
28 ft
28 ft
P = 112 ft A = 784 ft²
2)
18 ft
18 ft
18 ft
P = 54 ft A = 140.29 ft²
3)
27 ft
23 ft
P = 100 ft A = 621 ft²
4)
19 ft
19 ft
19 ft
P = 57 ft A = 156.31 ft²
5)
19 ft
19 ft
19 ft
P = 57 ft A = 156.31 ft²
6)
27 ft
27 ft
27 ft
P = 81 ft A = 315.67 ft²
7)
16 ft
19 ft
P = 70 ft A = 304 ft²
8)
16 ft
16 ft
16 ft
P = 48 ft A = 110.85 ft²
Page 76
1)
28 ft
28 ft
28 ft
P = 84 ft A = 339.49 ft²
2)
23 ft
21 ft
P = 88 ft A = 483 ft²
3)
26 ft
26 ft
26 ft
P = 78 ft A = 292.72 ft²
4)
17 ft
17 ft
17 ft
P = 51 ft A = 125.14 ft²
5)
16 ft
24 ft
P = 80 ft A = 384 ft²
6)
14 ft
14 ft
14 ft
P = 42 ft A = 84.87 ft²
7)
17 ft
18 ft
P = 70 ft A = 306 ft²
8)
22 ft
16 ft
P = 76 ft A = 352 ft²

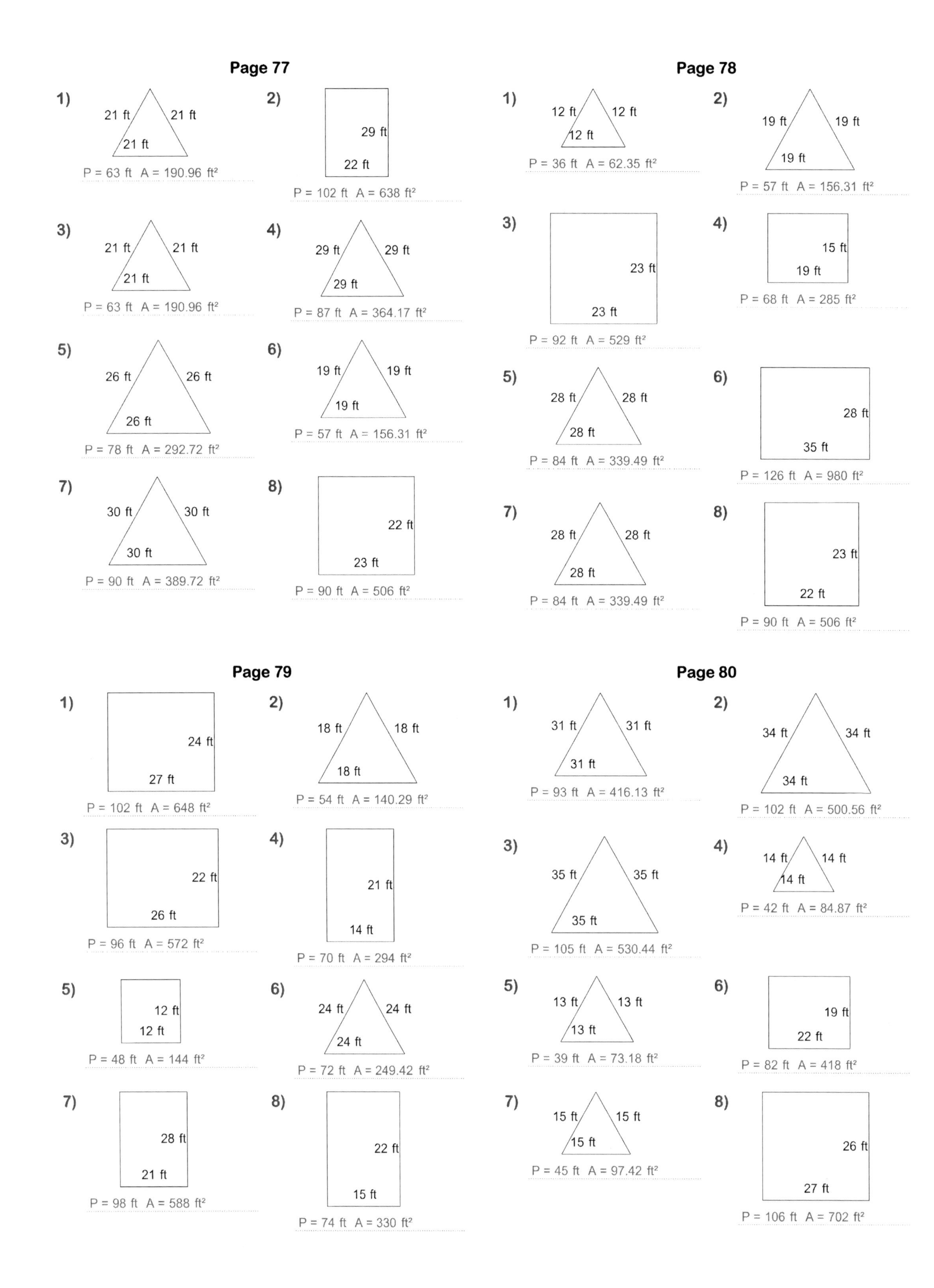
Page 77
1) 21 ft 21 ft 21 ft
P = 63 ft A = 190.96 ft²
2) 29 ft 22 ft
P = 102 ft A = 638 ft²
3) 21 ft 21 ft 21 ft
P = 63 ft A = 190.96 ft²
4) 29 ft 29 ft 29 ft
P = 87 ft A = 364.17 ft²
5) 26 ft 26 ft 26 ft
P = 78 ft A = 292.72 ft²
6) 19 ft 19 ft 19 ft
P = 57 ft A = 156.31 ft²
7) 30 ft 30 ft 30 ft
P = 90 ft A = 389.72 ft²
8) 22 ft 23 ft
P = 90 ft A = 506 ft²
Page 78
1) 12 ft 12 ft 12 ft
P = 36 ft A = 62.35 ft²
2) 19 ft 19 ft 19 ft
P = 57 ft A = 156.31 ft²
3) 23 ft 23 ft
P = 92 ft A = 529 ft²
4) 15 ft 19 ft
P = 68 ft A = 285 ft²
5) 28 ft 28 ft 28 ft
P = 84 ft A = 339.49 ft²
6) 28 ft 35 ft
P = 126 ft A = 980 ft²
7) 28 ft 28 ft 28 ft
P = 84 ft A = 339.49 ft²
8) 23 ft 22 ft
P = 90 ft A = 506 ft²
Page 79
1) 24 ft 27 ft
P = 102 ft A = 648 ft²
2) 18 ft 18 ft 18 ft
P = 54 ft A = 140.29 ft²
3) 22 ft 26 ft
P = 96 ft A = 572 ft²
4) 21 ft 14 ft
P = 70 ft A = 294 ft²
5) 12 ft 12 ft
P = 48 ft A = 144 ft²
6) 24 ft 24 ft 24 ft
P = 72 ft A = 249.42 ft²
7) 28 ft 21 ft
P = 98 ft A = 588 ft²
8) 22 ft 15 ft
P = 74 ft A = 330 ft²
Page 80
1) 31 ft 31 ft 31 ft
P = 93 ft A = 416.13 ft²
2) 34 ft 34 ft 34 ft
P = 102 ft A = 500.56 ft²
3) 35 ft 35 ft 35 ft
P = 105 ft A = 530.44 ft²
4) 14 ft 14 ft 14 ft
P = 42 ft A = 84.87 ft²
5) 13 ft 13 ft 13 ft
P = 39 ft A = 73.18 ft²
6) 19 ft 22 ft
P = 82 ft A = 418 ft²
7) 15 ft 15 ft 15 ft
P = 45 ft A = 97.42 ft²
8) 26 ft 27 ft
P = 106 ft A = 702 ft²

Made in United States
North Haven, CT
16 November 2024